Marmots: Social Behavior and Ecology

MARMOTS
Social Behavior and Ecology

David P. Barash

STANFORD UNIVERSITY PRESS
Stanford, California / 1989

Stanford University Press
Stanford, California
© 1989 by the Board of Trustees
of the Leland Stanford Junior University

Printed in the United States of America
CIP data appear at the end of the book

To Eva and Jacob, and to their successful dispersal

Contents

Foreword by John F. Eisenberg

When I first read David Barash's monograph on marmot social systems, in 1973, I found it to be a remarkable piece of work. He had skillfully integrated behavioral data and environmental measurements to make an excellent case for the manner in which life-history strategies might have evolved within this genus. This new book integrates the earlier research with later efforts as well as much unpublished information on our Eastern marmot *Marmota monax*, the woodchuck.

In my opinion, Barash and his work fall into the "heroic" era of behavioral research, and I do not exclude some of my own earlier efforts from this classification. I well remember the idealism of the 1950s, when we hoped that the study of animal behavior might help lead mankind to the recognition of universal truths and that these in turn could guide us in our personal life and in the orderly instruction of our students. Some of this youthful optimism still pervades our writings.

During his burst of publication activity in 1973, David Barash scored several important points; one outstanding thesis was that a species' life-history strategy is a compromise between the availability of resources and the cost of reproduction. By studying marmots, he was able to demonstrate that social organization is the outcome of demographic processes that in turn are controlled by primary productivity.

The diurnal ground squirrels have been the subject of concentrated research by a number of workers. At one point, some of us argued that since much of the marmot's life is spent underground, an attempt to induct a few subjects into captivity in a setting with artificial burrows could yield insight. This was a tactic I employed in the 1950s with great profit in my studies of kangaroo rats. But David Barash and others proved that—with patience and effort—a great deal can be learned in observing the above-ground life of these lovely creatures.

Preface

> We need another and a wiser and perhaps a
> more mystical concept of animals. Remote from
> universal nature, and living by complicated ar-
> tifice, man in civilization surveys the creatures
> through the glass of his knowledge and sees
> thereby a feather magnified and the whole image
> in distortion. We patronize them for their incom-
> pleteness, for their tragic fate of having taken
> form so far below ourselves. And therein we err,
> and greatly err. For the animal shall not be mea-
> sured by man. In a world older and more com-
> plete than ours they move finished and complete,
> gifted with extensions of the senses we have lost
> or never attained, living by voices we shall never
> hear. They are not brethren; they are not under-
> lings; they are other nations, caught with our-
> selves in the net of life and time, fellow prisoners
> of the splendour and travail of the earth.
>
> Henry Beston, *The Outermost House*

Many of us—even if we take ourselves very seriously as scientists, students of nature, or just observers—simply cannot shake the primitive
satisfaction that comes from looking into the lives of animals and then,
like foreign correspondents, reporting our observations of what Beston
called these "other nations."

My initial, and rather naive, expectation when I first began graduate
work in 1966 as a student of zoology professor John T. Emlen at the University of Wisconsin, was that I would study the behavior and ecology of
Olympic marmots for a master's thesis, and then "pull the genus together" in my Ph.D. research, somehow providing a coherent, synthetic
picture of the evolution of behavior among all marmots—that is, the
genus *Marmota*. As it often does, however, reality intervened. The Olym-

pic marmots alone occupied three years of field work, and I have been studying one marmot species or another every summer since then. And still, the genus isn't ready to be pulled together, at least not by me. Or rather, with an eye on Mr. Beston's advice, I should say that the marmots are entirely "together," and in need of no pulling at all. But try as I might to refrain from measuring animals by human standards, I simply have not been able to keep from seeking to understand them—"finished and complete" as they are—if only because, in the process, I have sought to help finish and complete myself.

The approach I take is sociobiology, the systematic study of the biological basis of social behavior. This newly emerged discipline bases its insights on evolution and the implications of natural selection for understanding behavior, notably social behavior. Much of biology is the study of phenotypes, the observable somatic characteristics of living things. Behavior is probably the most complex and confusing of all phenotypes, and social behavior is probably the most complex and confusing of all behaviors. Not surprisingly, therefore, students of social behavior can readily find themselves overwhelmed by the extraordinarily diverse, kaleidoscopic array of animal social behavior, as a result of which the study of social behavior can become merely the cataloging of "what animals do." By contrast, the sociobiologic approach emphasizes that living things are, in a sense, strategists, honed by many generations of natural selection to do things that tend to maximize their success in projecting copies of their genes into the future: that is, they tend to maximize their fitness. (Or, taking another perspective, the genes constituting any individual tend to compete with other, alternative genes for success in projecting copies of themselves, identical by descent, into the future via other bodies.) Either way, this approach offers the hope of a coherent and much-needed paradigm, pointing the way from "ethological stamp collecting" to programmatic science.

Sociobiology emerged from a synthesis of ethology, ecology, and evolutionary genetics during the late 1960s and early 1970s, receiving its "official" name nearly a decade later (Wilson, 1975). Further details on sociobiology can be found in Barash (1982), King's College Sociobiology Group (1982), or Krebs and Davies (1984), among many. The present book will not seek to explicate sociobiology; I shall have my hands quite full enough with marmots. Moreover, I shall pretty much restrict myself to marmots and marmots alone, at least for the bulk of what is to follow. This is not to disparage the excellent material, both theoretical and empirical, already available on the various relatives of the genus *Marmota*. Rather, it is testimony in part to the sheer volume of that fine work, as well as to my self-defined goal: to present the social behavior and ecology of marmots, *sensu stricto*. Only in the final chapter, where I seek to propose a general theory of marmot social evolution, will I range afield and consider other animals, especially the closely related ground squirrels and prairie dogs.

Every discipline, especially a new one that is based initially more on theory than on data, runs the risk of substituting the former for the latter. This may occasionally be true of practitioners of sociobiology, particularly since this particular discipline was born in reaction to the virtually atheoretical compilation of natural-history facts that seemed to precede the emergence of sociobiology. Ideally, any study of animal behavior, whether "sociobiologic" or not, will involve an optimum mix of theory and empiricism, although, in practice, it is easy to err on one side or another. On the one hand, the armchair naturalist, comfortably garbed in theory and armed with elegant but untried equations and antiseptic computer models, risks spinning irrelevant yarns; on the other hand, the ardent field biologist who knows only what he or she has seen is in danger of missing the forest for the trees.

Marmots are in many ways ideal subjects for the study of sociobiology, and for the study of behavioral evolution as well. When I first conceived the project, more than 20 years ago and before the sociobiologic approach had matured, I interpreted study of the "evolution of behavior" as largely the construction of plausible behavioral phylogenies, the sort of work that had been so effectively done by Konrad Lorenz, Niko Tinbergen, and, before them, Oscar Heinroth. My focus gradually changed, however, as it became clear that evolution could be used in a more active, dynamic way than by erecting hypothesized homologies. At the same time, sociobiology itself was gathering momentum, on the strength of a similar appreciation: the recognition that evolution, and especially evolution by natural selection, provided a coherent paradigm of enormous power in interpreting why animals do the things they do. As Henry Beston put it, they are "caught with ourselves in the net of life and time," and as such their behavior should be amenable to certain general principles—sexual selection, kin selection, intrasexual competition, life-history evolution, etc.—principles that in varying degree and detail are part of the "splendour and travail" of so many of the net's fellow prisoners.

Even with this conceptual refocusing, marmots have continued to be my favorite subjects for study, largely because of their potential for "predictive" as well as "correlational" studies. Given some above-zero heritability of the traits in question, if natural selection has acted upon the ancestors of living organisms such that they maximize their fitness in the environments in which they have been selected, then behavioral differences should result from differences in the selective regimes. And such regimes themselves should differ as a function of differences in the environments occupied. In comparing and contrasting the sociobiology of different marmot species, I shall emphasize the North American species: in particular the hoary, Olympic, and yellow-bellied marmots and the woodchuck, those with which I am most familiar and about which the most is known. Regrettably, sociobiologic research on the Palearctic forms has not been extensive, although I shall include reference to the Old World species wherever possible.

Marmots are ideal subjects for the study of free-living animals' behavior, whatever kind of science one chooses to "do." Unlike most mammals (which is to say, most rodents), marmots are diurnal, large enough to be easily observed and distinguished as individuals, and relatively insensitive to the presence of a human observer; and because they are sedentary, the hard-working marmoteer can be confident that a day spent in a marmot meadow will be rewarded with data. Marmots also have the good grace to hibernate during the academic year.

And finally, most marmots—with the exception of woodchucks—have a fine aesthetic sense, or, at least, behave as if they do. They inhabit meadows and talus rockpiles, many of them at or above timberline in some of the world's most beautiful mountains. There is simply no better place to spend a summer. To be sure, there are drawbacks, such as rain, fog, and snow, as well as freezing temperatures during what is supposed to be a respite from winter—at which time the little creatures retreat snugly into their burrows, leaving the intrepid marmoteer to face cold, wet disappointment, not to mention smug end-of-summer greetings from warm, dry, well-equipped laboratory colleagues, inquiring about the summer-long "vacation." Several times, I must confess, I have even decided to wrap up my marmotology career—but then came recollections of high peaks, crisp air, adorable furballs carousing on white snow or lush flowery meadows under brilliant blue skies, and I somehow realized, each time, that there were still a few important questions I had yet to ask my marmot friends.

And each time, they provided answers, although not always the ones I expected. Moreover, as I wrote this book I found that the questions so outnumber the answers that I could write another book—possibly a more interesting one at that—on what we do *not* know about marmot sociobiology. It would in any event be longer than this one. And whereas this effort is filled with equivocations, the other would be quite definite.

About one-half of this book consists of my own new data and inter-pretations, not before published, or data that I had previously presented but now reinterpret. Another one-fourth consists of work that I have already published. And the remaining one-fourth consists of the work of other marmoteers; in this material, the data are theirs and the interpreta-tions, unless specified otherwise, are mine.

My marmot research consists of the following: (1) three years of study of Olympic marmots (*Marmota olympus*) in Olympic National Park, Washington, 1967–69: 1,962 hours of observation; (2) one year of study of hoary marmots (*M. caligata*) in Glacier National Park, Montana, 1970: 344 hours; (3) one year of study of yellow-bellied marmots (*M. flavi-ventris*) in Rocky Mountain National Park, Colorado, 1971: 327 hours; (4) two laboratory studies of woodchucks (*M. monax*) in central New York State, 1972: 193 hours; (5) three years of springtime observations of free-living woodchucks in central New York State, 1971, 1972, and 1973 combined with one summer, 1972, and two autumns, 1971 and 1972: 994 hours; (6) one year of study of Alpine marmots (*M. marmota*) in Vanoise National Park, France, 1973: 312 hours; and (7) 14 years of study of hoary marmots in the Cascade Mountains, Washington State, 1974–86: 4,431 hours. Field biologists will recognize that the number of hours spent actually observing represents only a small fraction of the total time necessarily devoted to the field research, which also includes trapping time, waiting in bad weather, and just doing the maintenance necessary to keep alive. Some of the sample sizes reported here are distressingly small; this is especially true of my work on woodchucks, hardly any of which has previously been published, partly for this reason. The results, when *n* is small, must therefore be interpreted with caution. However, given the extreme difficulty of obtaining even these data, I do not feel especially apologetic, and would be delighted to defer to anyone with more, either now or in the future.

At least among the Olympic and hoary marmots, adults are readily identified as individuals, on the basis of their physical appearance alone. For scientific studies, however, confidence must be bolstered. I therefore live-trapped and permanently marked 247 individuals, by toe-clipping and/or sewing plastic streamers through their shoulder fascia. When the issue appears important, and I do not have complete confidence in the year-to-year identity of crucial individuals, I have stated this in the text; otherwise, I am confident of my data.

This preface would be incomplete if I did not point out that—at least for now—the accompanying book is my swan song to the marmots. In recent years, the appeals of marmot sociobiology have been eclipsed for me by a growing concern about the sociobiology of a different and far more difficult species, *Homo sapiens*, and its growing propensity to self-destruct. If and when that problem is resolved, however, or appears less acute, or the allure of marmoteering simply becomes too strong, I reserve the right to return to the marmots and their mountains.

In the meanwhile, thanks must go to the staff at the Academic Computing Center at the University of Washington, who labored (occasionally successfully) to retrieve data lost because of my pigheadedness and computer semiliteracy. I want especially to thank some of my graduate students, notably Warren Holmes, Bill Bernds, Suzanne Macy, Wendy Hill, Mike Hutchins, and Sam Wasser, for their enthusiasm, insights, and encouragement, and the various undergraduate assistants, who gave generously of their time and efforts. Dr. Robert Rausch kindly assisted with several translations from Russian; Preston Hardison helped prepare the illustrations. I also owe a great debt to my Soviet marmoteering counterparts who graciously made available numerous photographs of the Eurasian species, which have never before been published in the West, and to Dr. Robert S. Hoffmann, who facilitated the transfer. In several cases, unfortunately, the identity of the photographer was lost.

I am particularly grateful to Dr. Kenneth B. Armitage, whose long-term studies of yellow-bellied marmots provide a continuing source of valuable data and understanding, as well as an inspiring model of persistence and its benefits—even though we have not always agreed on interpretations. His work is cited so regularly in these pages that to me at least he almost feels like a co-author. Indeed, this book probably could not have been written without his research. I am also grateful to Dr. Armitage's various and productive students, notably J. Downhower, G. Svendsen, and D. Kilgore, for their studies of yellow-bellied marmots, without which my synthetic picture of marmot sociobiology, flawed though it may be, would be even more wanting. Ken Armitage and Jan Murie took time to read the manuscript in penultimate form and made numerous very helpful suggestions, which have greatly improved the final product. Judith Eve Lipton provided excitement, delight, consternation, and love, as ever.

At different times, and often more than once, my work was supported by the Society of the Sigma Xi, the Theodore Roosevelt Memorial Fund, the Olympic National Park Natural History Association, the Glacier National Park Natural History Association, the Research Foundation of the State University of New York, the Graduate School Research Fund of the University of Washington, the National Institute of Mental Health, the

Centre National de la Recherche Scientifique (France), the Harry Frank Guggenheim Foundation, and the National Science Foundation. I am grateful to all, but especially to the animals themselves, for helping me to achieve a more sociobiologic and perhaps even a wiser, if not a more mystical, concept of them, and also of myself.

D.P.B.

PART

I

Basic Biology

How much wood would a woodchuck chuck if a
woodchuck could chuck wood?

Anonymous

Every group of animals is unique, and marmots are no exception. Certainly, there is no reason to think that marmots are any more special than seagulls, bats, or bumblebees. But they do offer certain advantages to the student of animal behavior, especially to anyone interested in examining the evolutionary process for insights into general shapes and patterns. Thus, marmots are diurnal, large enough to be readily observed, and reasonably insensitive to a nearby marmoteer. They are also widely distributed throughout the Northern Hemisphere, occupying a variety of cool, often mountainous environments. They have received enough study that we can now say some coherent things about them, and, in the process, enough answerable questions have been raised to make the process rewarding, and also ongoing. This book seeks to document some of the important things we currently know about marmots, to point out what this teaches us about how the evolutionary process acts on social behavior, and to whet the appetites of future marmoteers.

CHAPTER I

Taxonomy, Physical Characteristics, Habitats, and Distribution

M armots are sciurid rodents, members of the tribe *Marmotini*, all members of one genus (*Marmota*), and without doubt closely related. Although they differ somewhat in size and pelage, marmots share a similar body shape and life style. Despite these similarities, however, they occupy a wide range of environments, raising the tantalizing possibility that differences in their basic biology and in their social behavior can be correlated in some meaningful way with variations in the environments within which they evolved and to which they are presumably adapted.

There are six species of marmots in North America, and another eight in Eurasia. Of these, the woodchuck (*Marmota monax*) is best known to most Americans. The name "woodchuck" is misleading, since marmots do not actually "chuck" wood, whatever that may mean. "Woodchuck" is probably an English corruption of the Cree *otcheck* and/or the Chippewa *otchig*, which in fact referred to the fisher (*Mustela fisheri*), a large arboreal weasel not closely related to the marmots at all. The other North American marmots are to varying degrees montane residents. The Old World species inhabit either the high plains—the "steppe"—or mountain slopes, as do their congeners in western North America.

Marmots are rodents, and rodents are the most abundant mammals, in sheer numbers of individuals as well as numbers of species. They range from the enormous capybaras of tropical South America (like scaled-down hippos) to the tiniest of mice. And they live just about anywhere on land. All rodents have two pairs of ever-growing, chisel-like incisors, which presumably hold the key to their success. Within the order Roden-

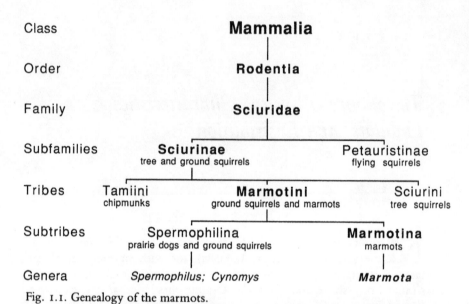

Fig. 1.1. Genealogy of the marmots.

tia, the squirrel family Sciuridae consists of more than 50 genera of medium to large animals. Among the New World species, the sciurids are generally divided into two subfamilies, the tree and ground squirrels, or Sciurinae, and the flying squirrels, or Petauristinae. The Sciurinae in turn are further divided into three major tribes, the Tamiini, or chipmunks (genera *Tamias* and *Eutamias*), the Sciurini, or tree squirrels (genera *Sciurus* and *Tamiasciurus*), and the Marmotini, or ground squirrels. Within the Marmotini, we can identify the subtribe Marmotina (the marmots, genus *Marmota*) and the subtribe Spermophilina (the prairie dogs, genus *Cynomys*, and the ground squirrels, genus *Spermophilus*, formerly *Citellus*; Hafner, 1984); see Fig. 1.1.

Marmots are the largest of the ground squirrels, although they are not the largest rodents, that honor going to the hystricomorphs, including the porcupine in North America and the capybara and its relatives. Marmots are also smaller than beavers (*Castor canadensis*), although their basic body shape (except for their tails!) is similar: they are large, heavy-bodied animals with relatively short legs, long bushy tails, a compact, sturdy physique, and short ears (Fig. 1.2).

The 14 identified marmot species, with their common and scientific names, are as follows:

NEW WORLD

(1) woodchuck or groundhog   *Marmota monax*
(2) yellow-bellied marmot   *M. flaviventris*
(3) hoary marmot   *M. caligata*
(4) Vancouver marmot   *M. vancouverensis*
(5) Olympic marmot   *M. olympus*
(6) Brower's, Brooks Range,
     or Arctic marmot   *M. broweri*

OLD WORLD

(7) black-capped marmot   *M. camtschatica*
(8) bobak or steppe marmot   *M. bobak*
(9) tarbagan or Mongolian marmot   *M. sibirica*
(10) Himalayan marmot   *M. himalayana*
(11) Menzbier's marmot   *M. menzbieri*
(12) long-tailed or red marmot   *M. caudata*
(13) gray or Altai marmot   *M. baibacina*
(14) Alpine marmot   *M. marmota*

Fig. 1.2. Adult male hoary marmot (*M. caligata*), showing characteristic pelage. (Photo by D. P. Barash)

Fig. 1.3. Worldwide geographic distribution of marmots; numbers correspond to list in text.

The worldwide distribution of these species is shown in Fig. 1.3. Experts differ over the exact number of species recognized (e.g., Rausch, 1953; Ellerman and Morrison-Scott, 1951), especially among the Asian forms; it has been suggested, for example, that *M. baibacina* and *M. sibirica* may hybridize in their zone of overlap in Mongolia (Smirin et al., 1985). No other cases of hybridization have been reported. There is general agreement on the distinctiveness of the North American species. Six of the eight Old World species have a diploid number of 38; only *M. camtschatica* has 40 (*M. himalayana* has not been karyotyped). Among the North American species, diploid numbers are as follows: *broweri*, 36 (Rausch and Rausch, 1965); *monax*, 38; *olympus*, 40; *vancouverensis, flaviventris*, and *caligata*, 42 (Rausch and Rausch, 1971). There also appear to be numerous morphological similarities between *M. camtschatica, M. broweri*, and the "*caligata* group"—*M. olympus, M. caligata*, and *M. vancouverensis* (Hoffmann, Koeppl, and Nadler, 1979)—possibly suggesting recent common ancestry.

The different marmot species vary somewhat in color but are generally shades of brown, typically with streaks, splotches, and flecks of yellow, gold, white, or black. Long, coarse guard hairs project several inches beyond the soft, woolly underfur. The guard hairs are commonly tipped with light buff or chestnut brown, giving most marmots a somewhat grizzled appearance (Fig. 1.4). One species, the Vancouver marmot (*M. vancouverensis*), found only on Canada's Vancouver Island, is melanistic (Fig. 1.5); melanistic individuals of *M. flaviventris* are also known to occur in the Teton region of Wyoming. Other species, notably the bobak marmot (*M. bobak*) of the western USSR, are quite pale, almost blond (Fig. 1.6). Marmots generally experience a single annual molt, beginning variably but sometimes as soon as emergence from hibernation. By midsummer, molting is far advanced in all individuals except the young of the year. At least one species, the Olympic marmot (*M. olympus*), molts twice: adults undergo a dark molt in midsummer that progressively covers the entire body, so that, by hibernation, they are nearly black. When they emerge the following spring, they are once again in yellow-brown pelage (Fig. 1.7). Molting patterns among woodchucks have been described in great detail (W. J. Hamilton, 1934). In the Old World bobak or steppe marmot, adults begin molting around mid-May, young of the year in mid-June (Shubin, Abelentsev, and Semikhatova, 1978). Pelage differentiation is sufficiently great among Olympic and hoary marmots to permit individual identification of most animals older than young of the year.

Marmots generally hibernate, for varying lengths of time depending on the severity of their winter environment and the occurrence of droughts during the summer. All marmots seem to undergo a characteristic pattern of weight gain and loss, gaining weight during the active season and then losing weight during hibernation (Fig. 1.8). The high-altitude Olympic and hoary marmots, whose hibernation period is longer, lose a higher

Fig. 1.4. Adult female woodchuck (*M. monax*), showing the typical grizzled appearance of the fur. (Photo by New York State Department of Environmental Conservation)

percentage of their body weight during hibernation than does the low-elevation woodchuck (Fig. 1.9). Among yellow-bellied marmots, weight loss during hibernation is also substantial, occasionally up to 50% between late fall and spring emergence (Armitage, Downhower, and Svendsen, 1976).

Adult males tend to be larger than adult females among all species of *Marmota* (Zimina, 1978, includes a review of the Old World species). However, the degree of sexual dimorphism varies among the different species, as does the seasonal pattern of weight gain, the age at which

Fig. 1.5. The Vancouver marmot (*M. vancouverensis*), a melanistic species. This individual is presumed to be a two-year-old. (Photo by V. Heinsalu-Burt)

Fig. 1.6. Adult female bobak marmot (*M. bobak*) from Soviet Kazakhstan, with three young of the year. (Photo by V. Maschkin)

Fig. 1.7. Adult male Olympic marmot (*M. olympus*) in late July, by which time its black molt is well established. (Photo by D. P. Barash)

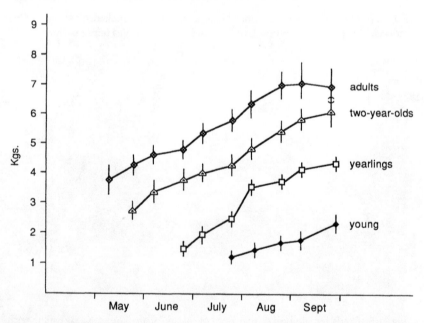

Fig. 1.8. Seasonal weights of *M. caligata*, averaging data for males and females for each age class. Based on 10 years of data and a minimum of two measurements for each cohort, including no fewer than 25 individuals in each category and 15 weighings for each month.

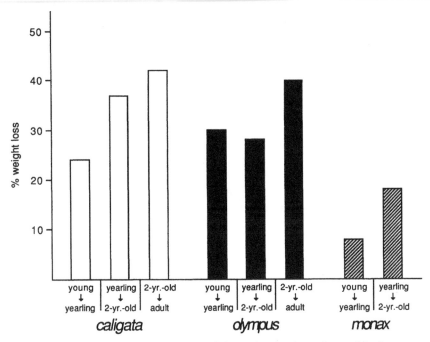

Fig. 1.9. Percentage weight loss during hibernation by *M. caligata*, *M. olympus*, and *M. monax*. Data for *caligata* based on the regime described for Fig. 1.8; data for *olympus* based on a minimum of eight individuals in each category; data for *monax* based on a minimum of nine individuals in each category.

growth stops, and the age at which sexual maturation is reached (Fig. 1.10). These issues will be developed, and explanations suggested, in subsequent chapters. Woodchucks appear to be unusual in continuing to lose weight for several weeks after spring emergence from hibernation, whereas the other species begin gaining weight promptly. This may be related to those species' longer hibernation period, such that they emerge with less fat available in reserve. (Pregnant females are an exception, since they do not begin gaining weight until after their young are weaned, and in some cases, adult males fail to gain or actually lose weight for a time after emergence.)

Marmots occupy a wide range of terrestrial habitats but avoid dense forest and swamp or marsh. Woodchucks are typically found in open, well-drained mesic fields, along hedgerows, adjacent to small woodlots, and near forest edges. Their habitats are generally flat or gently sloping, often include cultivated or abandoned cropland (Fig. 1.11), and are characterized by grasses, forbs, low shrubbery, and such deciduous trees as maple (*Acer*), elm (*Ulmus*), oak (*Quercus*), and cherry (*Prunus*). Woodchucks extend to northern and western North America, where they inhabit the low-elevation spruce/birch/aspen forests of east-central Alaska, typically along gravel beds and river valleys, as far as the head of Fortymile Creek in the eastern Alaskan lowlands. Their range extends eastward through central Canada past the southern border of Hudson Bay, south to northern Idaho in the west and northern Arkansas and Alabama in the east. In most of the conterminous United States, the western limit of woodchuck distribution is the Great Plains, although the species is best known from studies in the mid-Atlantic states.

The Asiatic *M. bobak* seems to prefer open steppe not unlike the habitat

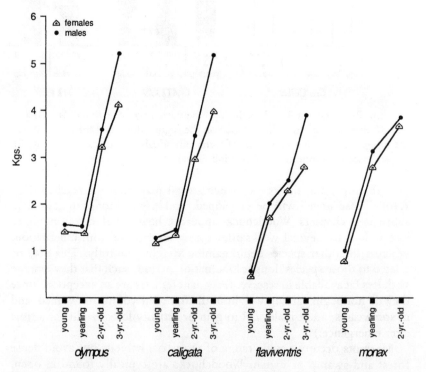

Fig. 1.10. Sexual dimorphism in weights of four marmot species. Data for young are within 2 weeks of weaning; for other age classes, mid-June weights are presented. *M. flaviventris* data based on Armitage, Downhower, and Svendsen (1976).

Fig. 1.11. Woodchuck (*M. monax*) in New York State hayfield. (Photo by New York State Department of Environmental Conservation)

of the North American black-tailed prairie dog (Formozov, 1966; Fig. 1.12). Like the prairie dog, *M. bobak* is apparently much less abundant today than in previous centuries, because of the clearing of grassland for agriculture (Kirikov, 1966; cited in Zimina and Gerasimov, 1973). In addition, the Eurasian species *M. bobak*, *M. baibacina*, and *M. caudata* have been implicated as natural carriers of plague (Zimina, 1978); as a result, eradication programs have greatly reduced their numbers, despite the fact that marmot pelts are valued as fur in the USSR. *M. himalayana*, on the other hand, remains quite common on flat, upland slopes of the Tibetan plateau, as well as in more rugged mountain terrain (G. Schaller, personal communication).

Most of the other species, both New World and Old, occupy a range of montane habitats, in which talus slopes, alpine and subalpine meadows, and rocky outcrops are prominent. After the woodchuck, the yellow-bellied marmot occupies the most diverse habitat of the North American forms, occurring in environments ranging from warm, rocky, xeric sites in the low deserts of eastern Oregon and Washington to lush subalpine and high alpine meadows in the Sierra Nevada and southern Rocky Mountains (Fig. 1.13). The vegetation in a well-studied and heavily populated region near Gothic, Colorado, is characteristic of the *Festuca thurberi*

Fig. 1.12. Two bobak marmots (*M. bobak*) looking out of their burrows in the Soviet steppe. (Photo by A. Nikolskii)

community (Svendsen, 1974; Langenheim, 1956), which is dominated by such grasses as *Festuca*, *Bromus*, and *Poa*, and forbs such as *Taraxacum*, *Senecio*, *Artemisia*, *Erigeron*, and *Viguiera*. Yellow-bellied marmots have even been reported occupying a riverside cottonwood (*Populus*) grove, where they moved vertically through the trees and sunned themselves horizontally on the limbs (Garrott and Jenni, 1978). Particularly toward the southern limits of its range, *M. flaviventris* tends to occupy isolated montane "islands."

Where yellow-bellied and hoary marmots are sympatric, for example, in the Purcell and Flint Creek Ranges in Montana, the former tend to be restricted to lower elevations (Hoffmann, 1974). In the absence of hoary marmots, yellow-bellied marmots occupy alpine habitats, as in the Beartooth Mountains of northern Wyoming and southern Montana. At the southern limit of its range, as in the Sangre de Cristo Mountains of New Mexico, *M. flaviventris* is found only in higher-elevation, alpine habitats (personal observations).

The hoary, Olympic, and Vancouver marmots, by contrast, are limited to alpine and subalpine mountain slopes (Douglas and Bliss, 1977) of the northwestern United States, western Canada, and Alaska. There has been little analysis of microhabitat preference among these marmots, although tall sedge communities, characterized by mesic grass and *Valeriana*, are especially important for Olympic marmots (Wood, 1973; Kuramoto and Bliss, 1970). Their lush subalpine meadows are characterized by sedges (*Carex*), numerous grasses, and forbs such as *Lupinus*, *Polygonum*, and *Erythronium*, bordered by clumps of alpine fir trees (*Abies*) with occasional krummholz Douglas fir (*Pseudotsuga*) and Western white pine (*Pinus*). Olympic marmots inhabit elevations between 1,700 and 2,000 meters, where winters are long, wet, and snowy, and summers are short, warm, and variable. Although temperatures there may occasionally ex-

Fig. 1.13. Two yearling yellow-bellied marmots (*M. flaviventris*) in Rocky Mountain National Park, Colorado. (Photo by D. P. Barash)

Fig. 1.14, *top.* Adult black-capped marmot (*M. camtschatica*) in a talus slope over-looking its permafrost habitat in northern Siberia. (Photo by unknown Soviet biologist)

Fig. 1.15, *bottom.* Black-capped marmots (*M. camtschatica*) from the Kamchatka Peninsula in northeastern USSR. This is the only known situation in which mar-mots live right by the ocean. (Photo by T. Lisitzyna)

ceed 28° C, and successive weeks without rain are not unknown, especially during August, snow may fall in any month.

Hoary marmots are found nearly at sea level in northern Alaska, but occupy higher elevations and talus slopes farther south. Hoary marmot environments are similar to those of Olympic marmots, varying from semi-arid alpine conditions, with such indicator vegetation as Lyall lupine (*Lupinus lyalli*) and pink plumes (*Sieversia campanulata*), to rich stream beds occupied by marigolds (*Caltha leptocephala*), anemones (*Anemone hudsoniana*), and yellow violets (*Viola glabella*). The Brower's marmot (*M. broweri*) of the Brooks Range in Alaska occupies similar habitat, although at higher latitude and lower elevation. This species is most abundant on damp soils, protected from the wind, in plant communities composed of *Saxifraga*, *Lupinus*, *Potentilla*, *Salix*, and various mosses and lichens (Loibl, 1983).

M. broweri and the Asiatic *M. camtschatica* and *M. sibirica* inhabit permafrost zones (Figs. 1.14 and 1.15). Zimina and Gerasimov (1973) suggested that the ancestral marmots were adapted to periglacial conditions, which facilitated their range expansion to Asia from North America, where they are generally acknowledged to have originated (Black, 1972). They may well have spread via "nunataks," vegetated terrain emerging from glaciers (Ives, 1974), after which populations became isolated and differentiated, as on the Olympic Mountains and Vancouver Island. Major rivers and densely forested lowland valleys constitute substantial barriers to marmot dispersal, so their mountain habitat resembles a series of islands in a ocean of inhospitable, low-elevation terrain. Although American zoologists have considered the North American woodchuck to be similar to the primitive marmot from which others evolved and differentiated, it is also possible that it is a postglacial specialist, and that the primitive marmots resembled the currently more abundant montane forms. In any event, late Pliocene fossil marmots found in the western USSR show that the genus emigrated to the Old World before the beginning of the Pleistocene (Zimina and Gerasimov, 1973).

The habitat preferences of several Asiatic marmot species are described by Zimina and Gerasimov (1973) as follows:

In the high mountains of the Tien Shan and Pamirs, *M. baibacina* and *M. caudata* inhabit alpine and subalpine meadows, meadow-steppe, and the peculiar cold steppes, semideserts and deserts in the eastern Pamirs (Zimina, 1958). Fine-grained, soft soils are preferred for burrowing, but marmots also inhabit areas of massive accumulation of large boulders and screes, stony deltas of ancient streams, and, most often, broken rock or pebbles with fine-grained deposits. . . .

Fig. 1.16. Gray marmot (*M. baibacina*) from Soviet Kazakhstan. (Photo by un-known Soviet biologist)

In the regions of eastern Siberia and Kamchatka, *M. camtschatica* lives in even more severe conditions. This species inhabits forestless, alpine areas termed *goltzi*, and occupies enormous areas of uplands and mountain ridges. As well as broken rocky soils and scanty grass, these *goltzi* have permafrost underlying most of their area, although at a lower level than in bog soils. In addition to *goltzi*, *M. camtschatica* inhabits warm, south-facing, meadow-steppe slopes (especially in the Transbaikal region), alpine meadows with elfin wood of Siberian dwarf pine (*Pinus pumila*), and the summits and ocean-facing slopes of volcanic mountains in Kamchatka. According to Kapitonov (1960, 1963), the population density of *M. camtschatica* is much lower than for all other species.

Environments of *M. bobak* and *M. baibacina* differ from those described above. In the hills of Kazakhstan, *M. baibacina* occupies only the highest mountain ridges and especially shallow valleys therein with mesic-steppe vegetation that retains its nutritional qualities for a long time (Kapitonov, 1966). The lower boundary of its distribution is 800 meters above sea level. In contrast, *M. bobak* dwells only in the northern part of the Kazakhstan hills, on lower, more gentle slopes. This marmot inhabits flatter areas, which are only moderately mesic and support steppe vegetation differing from the meadow-steppe occupied by *M. baibacina* [Fig. 1.16]. . . . The alpine marmot, *M. marmota*, presently inhabits the alpine and mountain-meadow zones of the Alps and western Carpathians [and also the Pyrenees]. This marmot lives chiefly in old morainal landscapes although it can be found within the mountain-forest zone, occupying a different kind of grassland on the karst plateau of the High Tatras in the Carpathians. [Bracketed material mine]

It is widely acknowledged that *M. monax* is more abundant now than when North America was first colonized by Europeans, largely because of the clearing of old-growth forest. Data are unavailable for the montane North American species, although both hoary and yellow-bellied marmots are abundant and seem likely to have maintained their populations since colonial times, possibly even to have increased. The population of Olympic marmots is low, probably not more than 2,000, but this species is almost entirely protected within Olympic National Park in Washington State, and there is no sign of a population crisis. The Vancouver marmot is the rarest species of North American marmot and is considered endangered by the Committee on the Status of Endangered Wildlife in Canada (unpublished report, Ottawa). The Brower's marmot, from the Alaskan North Slope, is virtually unknown and seems unlikely to have a large population. The status of the various Old World species is unclear, although most do not currently appear at significant risk; *M. menzbieri*, because of its unusually small range, may be an exception.

Summary and Conclusions

1. There are 14 species of marmots worldwide, of which six reside in North America. Their body shape is quite uniform, although weights vary among species and between sexes.

2. The woodchuck is a resident of low-elevation fields and forest ecotones; the other species live primarily in various higher-elevation habitats, including the Palearctic steppes and subalpine to alpine meadows.

3. Most marmot species are adapted to periglacial conditions, and they may well have spread geographically in association with glaciations, later to become isolated and differentiated.

CHAPTER 2

Seasonal Patterns and Hibernation

M armots typically hibernate during the winter and are active during the spring, summer, and early autumn. For example, Olympic and hoary marmots emerge during mid-May, when their meadows are generally covered by up to several meters of snow. They become increasingly lethargic by late August, and re-enter their burrows (immerge) for hibernation again beginning in early September, so that by October virtually all animals have retired for the year. In the Colorado Rockies at 2,900 meters elevation, yellow-bellied marmots are usually active from the first week of May until mid-September. Populations at lower latitudes and/or elevations tend to emerge from hibernation earlier. For example, at 1,700–2,200 meters in the Sierra Nevada of California, *M. flaviventris* emerges in early April (Nee, 1969).

The following seasonal pattern (from Davidson et al., 1978) occurs in three populations of the long-tailed marmot (*M. caudata*):

Life event	Pamirs (4,000 m)	Gissar Mtns (3,000–3,500 m)	Turkestan (1,500–2,000 m)
Emergence	April 15–30	March 21–31	March 1–15
Birth	June 27–July 7	May 22–June 9	no data
Hibernation	September 16–27	August 25–28	July 25–August 7

In arid climates, immergence is typically early; hibernation (or estivation) may begin by midsummer, apparently a result of forage desiccation. Thus, yellow-bellied marmots inhabiting the deserts of eastern Washington sometimes immerge as early as June (Couch, 1930), and most older animals are inactive by mid-July (personal observations). A similar correlation has been reported for the Asian bobak marmot in central Kazakhstan, USSR (Shubin, 1963).

Latitudinal trends in seasonal emergence and length of hibernation are most apparent among woodchucks, since they do not vary significantly in the elevation of occupied habitats. Thus, in southern Illinois *M. monax* sometimes does not hibernate at all, and for individuals that do, torpor may last only a month or so (Anthony, 1962). In southern Ontario, most woodchucks begin hibernating by early September, and all have done so by October 1 (de Vos and Gillespie, 1960). These animals typically emerge during the first or second week of March, whereas in central New York State, a few adult male woodchucks emerge during the third week of February, with the rest of the population following suit by mid-March (W. J. Hamilton, 1934). Adult females emerge about 3 weeks later, and yearlings a week or so later yet. This pattern repeats itself among the montane marmots, but with the timing compressed because of the shorter season of above-ground activity.

Woodchuck emergence dates vary approximately 1 day for 16 km, or about 7 minutes of latitude (D. E. Davis, 1967a). In central Pennsylvania, adult males emerge typically during the first week of February . . . although not precisely on February 2, Groundhog Day. (Marmots, incidentally, are the only U.S. animal to have a day named after them. The legend of Groundhog Day apparently began about 200 years ago among the early Pennsylvania Dutch near Lancaster, Pennsylvania: an Old World tradition that badgers emerge from their burrows on Candlemas Day was transferred to the New World woodchucks [Reichman, 1942]; badgers are not found in the eastern U.S.)

Higher-elevation and/or higher-latitude populations tend to emerge later, although the relative sequence is unchanged. Different marmot species also differ in the variability of their seasonal patterns: the emergence dates of young *M. monax* show greater variance than do the emergence dates of young *M. flaviventris*, which, in turn, show greater variance than the emergence dates of young *M. caligata*.[1] This trend is consonant with the progressively more restricted period of seasonal activity in the sequence from *monax* to *flaviventris* to *caligata*. In addition, as we shall see, variability in these matters is compounded by species-to-species variability in the age at first breeding.

Within a few days to a few weeks after emergence, marmots excavate

[1]F-max test, P < .05, when emergence dates are transformed into days 1 through n, where 1 = the first date of emergence and n = the last observed; based on known emergence dates for 25 woodchucks, 16 yellow-bellied marmots, and 31 hoary marmots

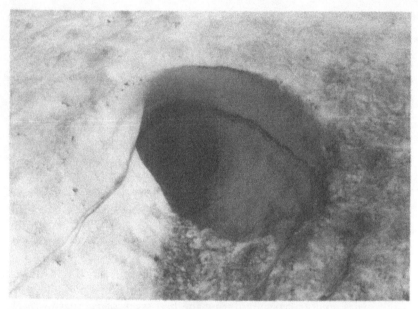

Fig. 2.1. Olympic marmot (*M. olympus*) burrow opening in late May. Like many others at this time of year, this burrow is covered by nearly 3 meters of snow. (Photo by D. P. Barash)

existing burrows, typically digging down through the snow and exposing the previous year's burrow openings (Fig. 2.1). This has been reported for woodchucks (de Vos and Gillespie, 1960), and I have observed it among both Olympic and hoary marmots. It is unknown whether such behavior is guided by sensory input or a remembered cognitive map of the terrain.

The Asiatic *M. camtschatica* (Kapitonov, 1960), *M. bobak* (Bibikov, 1967), and *M. caudata* (Davidov et al., 1978) breed within their burrows several weeks before emerging (Shubin, 1962). In these cases, therefore, the young of the year emerge only about 1 or 2 weeks after their parents are first seen above ground. However, for the New World species as well as the European *M. marmota*, breeding takes place shortly after the adult females emerge. Since gestation requires about 30 days and weaning at least another 30, young marmots in such cases first emerge about 2 months after their mothers. In central Pennsylvania, for example, adult female woodchucks are inseminated in mid-March, give birth in mid-April, and their young emerge by late May (Snyder and Christian, 1960). Among certain species, notably *M. bobak*, lactation is reported to continue for no

less than 50–55 days (Shubin, Abelentsev, and Semikhatova, 1978), resulting in a longer delay between breeding and emergence of the young
of the year.

Young yellow-bellied marmots are weaned between mid-June and early
August, typically at the end of June and in the first half of July (Armitage,
Downhower, and Svendsen, 1976). By contrast, young Olympic and
hoary marmots do not appear before the third week of July and, most
commonly, a week after that. Thus, the reproductive cycle of *M. monax*
(from copulation to weaning) occupies about 17 weeks; of *M. flaviventris*,
about 13 weeks; and of *M. olympus* and *M. caligata*, about 10 weeks.

Turning now to hibernation, there is an enormous literature on mammalian hibernation in general, and marmot hibernation in particular (see
D. E. Davis, 1976, and Lyman et al., 1982, for reviews). The annual
cycle consists of weight gain during the active, above-ground period followed by dramatic weight loss during hibernation. This loss may be
largely endogenous and also directly related to the onset of hibernation
itself. Thus, the rate of weight gain among woodchucks changes seasonally even with constant laboratory feeding (D. E. Davis, 1967b).
Weight gain is most rapid from April to June and markedly lower from
June to October; this cycle is not appreciably influenced by temperature,
day length, molting, or change of lipid content in the diet. It correlates
closely with the quantity of food consumed. Woodchucks maintained
with constant food, temperature, and photoperiod do not hibernate;
however, they do continue to show an annual cycle of weight changes,
thereby indicating that the food-consumption cycle is not "set" by hibernation. Castrated males undergo similar annual changes, suggesting that
gonadal function is not responsible (D. E. Davis, 1967b). However, environmental cues appear also to be involved, since within two years the
annual cycles of woodchucks shipped to Australia became entrained from
a weight maximum in October to a maximum in April (D. E. Davis,
1976).

It appears that under natural conditions, hibernation is induced by
food deprivation and is facilitated by body condition, notably the presence of large amounts of fat (Lyman and Dawe, 1960). Although woodchucks maintained with food and warm temperatures will not hibernate,
when kept at low temperature and without food they will begin hibernating at any month. Whereas food deprivation must last 3 to 4 weeks in
order to induce hibernation in summer, 3 to 4 hours may be sufficient in
winter (D. E. Davis, 1976), once again implicating endogenous timing.

Cycles comparable to that of *M. monax* have been demonstrated for
M. camtschatica (Filonov, 1961) and for captive *M. flaviventris* (Hock,

1969), although the latter will hibernate even under constant conditions, provided that disruptions are strictly avoided (K. B. Armitage, personal communication). Free-living Olympic and hoary marmots undergo a comparable body cycle, generally doubling their weight during the brief summer season and then losing at least a third during each hibernation. Free-living woodchucks have been reported to continue losing weight for several weeks following emergence (Snyder, Davis, and Christian, 1961), although since the same individuals were not sampled over time, the observed decrease could be due to later emergers being lighter than earlier emergers. Olympic and hoary marmots begin gaining weight almost immediately, although I have witnessed considerable annual variation, apparently as a function of snow cover at emergence. In years of heavy snow, most individuals—especially adult males—lose weight for up to several weeks, until new vegetation has sprouted. The Asiatic *M. caudata* apparently begins fat deposition in July, after first following the woodchuck pattern and losing weight for up to 2 weeks after seasonal emergence (Kizilov and Semenova, 1967).

There also appears to be considerable local ecotypic differentiation in patterns of weight gain, appropriate to differing environmental regimes. Thus, the annual rhythm of weight gain by captive Ontario woodchucks—which undergo a lengthy hibernation of 6 full months—involved significantly greater fluctuations than that of woodchucks captured in Virginia (D. E. Davis, 1971), where hibernation rarely exceeds 2 months. Captive yellow-bellied marmots also reveal circumannual cycles in food consumption, body mass, and metabolic rate which appear to be adaptations to their local environments. Animals living in the xeric lowlands of the Columbia River basin of eastern Washington State normally emerge in late February or early March; by contrast, animals from the Colorado Rockies do not emerge until early May. When animals from these two populations were maintained under constant laboratory conditions, the lowland animals showed a peak in food consumption at least two months before the montane animals did (Ward and Armitage, 1981).

Adaptive differentiation is also apparent in metabolic function: at low temperatures (5 to 10° C) montane animals had lower metabolic rates than did lowland individuals, whereas above 25° C, the metabolism of montane individuals increased with ambient temperature but that of lowland individuals did not (Ward and Armitage, 1981a). Such reduced metabolic rates at low temperatures may well be an adaptation for conserving energy at those temperatures that are more often encountered by montane animals; similarly, the reduced metabolism at high temperatures of individuals living in semi-arid habitats probably helps conserve water. The role of water stress in inducing immergence and hibernation among ground squirrels generally deserves more attention (Bintz, 1984).

When maintained at room temperatures and given food *ad libitum*, woodchucks show a seasonally cyclic metabolic rate, as measured by carbon dioxide production. Their metabolic rate peaks in May, when their fat level is lowest, and then declines until autumn (E. D. Bailey, 1965a). This seasonal decrease in metabolism may well facilitate weight gain, thereby creating a physiologic predisposition for hibernation, and also allowing continued fat deposition without great increases in food intake. It may well be a valid generalization that all marmots lay on fat during the late summer largely by reducing their metabolic rates. In captive *M. flaviventris*, for example, body mass increases from August to September, while actual food consumption *decreases* by 50% (Ward and Armitage, 1981a). Metabolic rates also decrease, suggesting that the critical parameter is energy balance rather than simple caloric input.

As summer proceeds, both hoary and Olympic marmots become noticeably more lethargic, although they continue to gain weight: eight yearling hoary marmots traversed an estimated 161 meters/day during the first 2 weeks in July (97 observation-hours) as opposed to an estimated 61 meters/day during the last 2 weeks in August (80 obs-hrs).[2] These same animals averaged a weight gain of 31.4 grams/day in July as compared to 27.8 grams/day in August (this difference is not statistically significant). During the first 2 weeks of July, the yearlings spent 31% of their above-ground time actively foraging; by the second 2 weeks of August, this proportion had dropped to only 12%.[3] Thus, weight gain continues almost unabated in late summer, although foraging is significantly reduced, as are levels of general activity, measured by linear distance traversed per day. Social activity is similarly reduced by this time (see Chapter 6).

Even when not preparing to hibernate, marmots generally have a low metabolism, which permits a high proportion of assimilated energy to be used in production rather than maintenance. This also appears to be an appropriate adaptation for a hibernator that must accumulate adequate fat if it is to survive the winter. Thus, patterns of energy flow in marmots apparently differ substantially from those found in nonhibernators. For example, tissue growth efficiency among active yellow-bellied marmots averages nearly 17%, about five times that of most endotherms. The

[2]t test, $P < .05$
[3]binomial test, $P < .05$

ratio of production to maintenance is nearly 30%, whereas a ratio of 1.1 to 3% is typical of endotherms (up to 42.8% for ectotherms). Although digestion efficiency among marmots is comparable to that of other homeotherms, marmots use only about 77% of their assimilated energy in maintenance, whereas poikilotherms must expend about 98% (Kilgore and Armitage, 1978). These findings underscore the importance of energy balance and adequate nutrient intake in the daily life of marmots.

In addition to their reproductive activities and the associated patterns of social behavior, weight gain—notably fat deposition—is a prominent aspect of marmot biology, and an essential one if individuals are to be successful. The proportion of fat accumulated by the end of the active season can be immense, and is apparently necessary for surviving until the next spring. For example, in early fall, white fat accounted for 1,155 grams out of one male woodchuck's total weight of 6,125 grams (Snyder, Davis, and Christian, 1961). This fat accumulation is almost certainly related to the probability of overwinter survival. Among both hoary and yellow-bellied marmots, young of the year that emerge earlier tend to

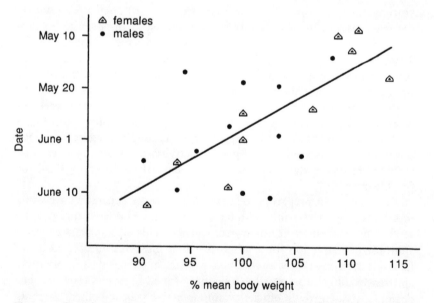

Fig. 2.2. Correlations between the weight of adult *M. caligata* and the date of snowmelt. Adult weight is presented as percentage of (10-year) mean body weight attained by adults by June 15. Date of snowmelt is when an estimated 50% of the meadow area is free of snow.

Fig. 2.3. Young tarbagan (*M. sibirica*). (Photo by unknown Soviet biologist)

weigh more at hibernation and to have higher overwinter survival (see Chapter 11). For adults of both species, body weight in June is strongly influenced by the date of the last accumulated snowfall; that is, when meadow vegetation is exposed early in the spring, marmots begin foraging earlier and, as a result, they gain more weight (Fig. 2.2).[4]

When tarbagans (*M. sibirica*) have gained insufficient fat because of poor spring and summer foraging conditions, either they begin hibernating later, or these skinny individuals tend to awaken early the following spring, before enough snow has melted to permit weight gain, whereupon they die (Kalabukhov, 1960; Fig. 2.3). Similarly, bobak marmots living near grazing livestock or near roads typically begin hibernating up to 4 weeks later than those living under natural conditions (Shubin, Abelentsev, and Semikhatova, 1978); this suggests that interruption of foraging results in delayed fat accumulation, which in turn generates a delay in the onset of hibernation. By contrast, when food is abundant, hibernation begins relatively early in the autumn or late summer. The steppe-dwelling *M. bobak* typically estivates in late summer: when sufficient rainfall results in good foraging conditions, torpor may begin as early as July, whereas in years of drought and consequent poor forage

[4] Spearman rank test for first date of more than 10% foraging and weight gain, $P < .05$

and diminished weight gain, it may not begin until early September (Shubin, 1963).

It might seem advantageous for marmots to emerge and breed as early as possible, so as to give their young a maximum head start and thereby enhance their prospects of success. But there is another consideration: hibernation is much more efficient than arousal; the former uses only about one-seventh the metabolic energy of the latter (E. D. Bailey and Davis, 1965). As a result, emerging marmots must calibrate their arousal, emerging early enough to allow their offspring a good chance of eventual success, but not so early that the adults wind up "turning on their engines" too soon and metabolizing their scanty springtime body resources before sufficient vegetation is available to replenish fat reserves that have been depleted by hibernation.

Summary and Conclusions

1. Marmots typically hibernate during the winter and are active during the spring, summer, and early fall.

2. There is considerable interspecies and intraspecies variation in the dates of hibernation and emergence, correlated with environmental severity.

3. High-latitude or high-altitude populations show reduced seasonal variability in their annual cycles.

4. Hibernation appears to result from a predominantly internal rhythm of weight gain and fat deposition, which varies interspecifically as well as intraspecifically.

5. Reduced metabolic rates and activity typically precede hibernation torpor and are associated with pre-hibernation weight gain; fat deposition is a prominent aspect of marmot above-ground activity and is apparently related to overwinter survival.

CHAPTER 3

Daily Activity Cycles

J ust as their seasonal patterns follow a predictable course, marmots also demonstrate consistent diel cycles in their behavior. I compiled activity budgets for Olympic marmots by censusing each animal at three different colonies every 2 minutes (Barash, 1973a) and categorizing behavior as either in burrow, looking out of burrow (either from within the burrow with only the head looking out, or seated or lying on the "dirt porch" just outside the burrow), feeding, social (greeting, play-fighting, etc.; see Chapter 6), miscellaneous (up-alert, carrying grass, moving, etc.), or unknown (if the animal was not visible at the time of the observation). About 30 point-censuses were obtained for each animal per hour. These results indicate the percentage of time spent by each age and sex class in these basic activities.

Figure 3.1 presents a July activity budget obtained in this manner for *M. olympus.* Yearlings generally emerge during mid-June, about one month later than the adults. The basic pattern is shown by all animals, with distinct morning and afternoon activity periods, most prominent in July. In addition, age and sex classes are distinguished in different months. Although yearlings and adult females in June spend relatively more time in burrows between 1300 and 1600 hrs than at other times, adult males do not.[1] Adult males spend significantly more time looking out (presumably alert) than do adult females or yearlings.[2] This difference remains statistically significant during May, June, and July, but fails to hold during August. The generally higher early-season activity levels of males reflect their greater wandering and social involvement at this time (see Chapter 8).

[1] binomial tests, P < .05
[2] binomial tests, P < .05

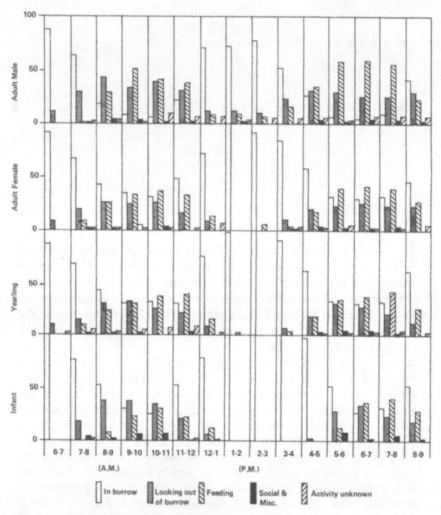

Fig. 3.1. Activity budgets for the Olympic marmot (*M. olympus*) during July. Based on 148 hours of observation; ordinate represents the percentage of time spent in each activity.

By July, activity is distinctly bimodal, with a midday activity lull obvious for all animals; between 1200 and 1600 hrs significantly more time is spent in the burrow and less time is spent in all other behaviors.[3] The

[3]binomial tests, P < .05

frequency of looking out by adult males in July is not significantly higher than for other animals, although adult males' early morning levels are the highest of all animals'. From 1200 to 1600 hrs, adult male above-ground activity significantly exceeds that of all other groups,[4] indicating that males are less subject to the midday activity lull. Males' feeding frequency is significantly higher in July than in June,[5] while for yearling and adult females the frequency of feeding is statistically indistinguishable during the two months. Intense male social behavior during June probably gives males less time for feeding than during July, while other animals are less affected. The sociality of adult males and of young of the year is very high immediately after emergence, but this is not reflected in their activity budgets, largely because social interactions are generally brief and often not revealed by a point-census.

By August, the midday activity lull is still apparent but is reduced in all animals except the young, which remain inactive during the early afternoons. The morning and afternoon activity periods of all other animals are reduced from the levels of July: yearlings, adult females, and adult males all spend significantly more time in their burrows during August.[6]

The activity budget of Fig. 3.1 was compiled during fair weather. During bad weather (fog, rain, snow), things are quite different. Figure 3.2 presents an activity budget recorded at a single *M. olympus* colony during three consecutive July days of heavy fog and nearly constant rain. No midday activity lull occurred, and activity began later and ended earlier in the day. Distance traveled from the home burrow was also greatly restricted in such weather: during three days of clear weather in July, four adult males averaged an estimated 225 meters of linear travel per animal per day; by contrast, during the three days presented above, two adult males averaged 33 meters per animal per day.[7]

Early season post-emergence activity patterns are also distinctive. Figure 3.3 reports activity budgets of a single *M. olympus* colony recorded on 5 consecutive days (May 8 to 12), beginning at the initial day of emergence from hibernation. The animals remained within 5 meters of the home burrow at all times, engaged in occasional social interactions, and briefly ate snow. Activity began late, ended early, and there was no midday reduction. Essentially, the marmots spent their time either in their burrows or looking out, doing virtually nothing else. Within about

[4] binomial tests, $P < .05$
[5] binomial test, $P < .05$
[6] binomial tests, $P < .05$
[7] Mann-Whitney U test, $P < .01$

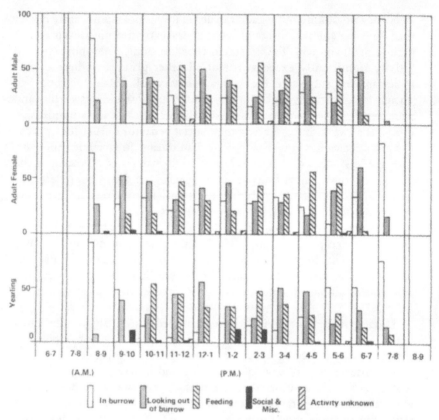

Fig. 3.2. Activity budgets for the Olympic marmot (*M. olympus*) during light rain and heavy fog (July 18–20). Based on 44 hours of observation; data presented as in Fig. 3.1.

2 weeks, social interactions increased, and the animals moved several hundred meters to snow-free areas where they began foraging.

Olympic marmots may have to relearn the physical landmarks of their colony each spring. Two individuals that I trapped on the second day after emergence and released about 7 meters from their burrow appeared disoriented and walked slowly and aimlessly for about 15 minutes, after which they flattened themselves against the snow and remained motionless for another 15 minutes. I saw this behavior and posture at no other time. I then chased these animals back to their burrows; they moved reluctantly until they were within a few meters of the burrows, when they suddenly showed distinct orientation and quickly ran in. Following their

release later in the season, by contrast, trapped animals always run un-
erringly to the nearest burrow, from distances of up to about 15 meters.
The restricted movement normally observed immediately after spring
emergence may therefore occur because the animals have "forgotten" the
colony landscape, abetted by the fact that snow typically obscures colony
landmarks early in the spring.

The onset of hibernation is gradual, with progressive but often erratic
decrement in social activity, foraging, and time spent above ground.
Both Olympic and hoary marmots occasionally remain underground for
several consecutive days in early September, emerging only briefly, with
little foraging. At such times, their activity is almost exclusively limited
to looking lethargically out of a burrow or sitting just outside. Adult
males and nonparous females become inactive earliest, while yearlings,
young of the year, and parous females remain active longest, presumably
because of their greater need for weight gain before hibernating (see
Chapter 12).

The above patterns, with their seasonal shifts, appear characteristic

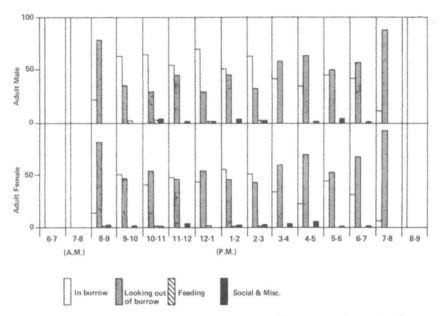

Fig. 3.3. Activity budgets for the Olympic marmot (*M. olympus*) during the first
5 days after emergence one year. Based on 51 hours of observation May 8–12;
data presented as in Fig. 3.1.

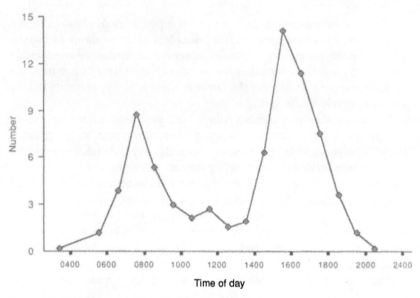

Fig. 3.4. Mean number of free-living woodchucks (*M. monax*) observed per hour in south-central Pennsylvania. Based on data from 544 woodchucks observed between June 16 and August 15 in two consecutive years. (Data from F. H. Bronson, 1962)

of all marmots. The daily activity pattern of *M. caligata* is virtually indistinguishable from that of *M. olympus*. *M. flaviventris* also shows a bimodal activity peak during midsummer, with unimodal tendencies early and late in the season (Travis and Armitage, 1972; Armitage, 1965). Adult male yellow-bellied marmots are the most active animals during the first few weeks after emergence: they are the first out of their burrows each day and the last in. Several weeks after emergence, however, adult females become the most active (Armitage, 1965). In *M. monax* as well, above-ground activity is most pronounced during midday in the spring, followed by morning and late afternoons during the summer (Meier, 1985; F. H. Bronson, 1962; Fig. 3.4), then midday once again in the autumn (Grizzell, 1955), often shifting toward late afternoon just prior to hibernation (de Vos and Gillespie, 1960). Captive Brower's marmots, maintained under laboratory conditions in Maryland, also showed a characteristic bimodal daily activity cycle (Loibl, 1983).

The diel cycle may be strongly influenced by temperature constraints. Captive *M. flaviventris* maintained out of doors showed a rectal temperature rhythm with an average of 37.4° C, and maxima reached between

1100 and 1600 hrs, when the animals are normally underground (Pattie, 1967). Many hibernators are only imperfect endotherms during their active periods, and marmots appear to be consistent with this tendency. Radio telemetric monitoring of free-living yellow-bellied marmots showed a peak in body temperature around sunset, followed by a substantial drop of nearly 2° C (Downhower and Pauley, 1970). Marmots have a relatively low weight-specific metabolic rate (Kilgore, 1972), which permits them to store a high proportion of the energy ingested, as we have seen. Marmots may facilitate energy intake by adjusting their behavior to capture radiant energy during periods of low temperature: *M. olympus, caligata*, and *monax* all tend to orient their bodies broadside toward the sun, especially in the early morning when the animals sprawl on the ground or rocks near the entrance to their burrows. A similar pattern has also been reported for *M. flaviventris* (Travis and Armitage, 1972; Figs. 3.5 and 3.6).

On cloudy days, not surprisingly, marmots do not orient in this manner, and they spend less time lying about near their burrow entrances. Among hoary marmots, for example, point-censuses of 38 different individuals every 5 minutes revealed that during clear days in July, animals spent 44% of their above-ground time in the early morning sunning themselves, presumably thereby boosting their thermal input, whereas during cloudy days, less than 10% of their time is spent in comparable postures.[8] On clear days, they spend less than 12% of their time between 0600 and 0900 hrs foraging, whereas on cloudy days, they are foraging nearly 25% of that time.[9] In the former case, air temperature is generally higher than body temperature, and it may well be that marmots sun themselves to raise their internal temperature; in the latter case, air temperature is generally lower than body temperature, so marmots may well switch to foraging when it is more profitable energetically than sunning.

During hot weather, avoidance of midday activity may well reflect avoidance of midday sun and consequent thermal stress. Foraging activity among yellow-bellied marmots is apparently not a function of ambient temperature, however, so long as it is below 25° C, a "critical temperature" at which yellow-bellied marmots experience their minimum metabolic rate (Kilgore, 1972). Above 25° C, thermal stress may be significant, restricting options for above-ground activity (Webb, 1981), and causing retreat into cool burrows. This number may be misleading, however (K. B. Armitage, personal communication), because the crucial factor is more likely the animal's actual operating temperature, which

[8] binomial test, P < .05
[9] binomial test, P < .05; 17 individuals on clear days and 11 individuals on cloudy days

Fig. 3.5, *top*. Adult female yellow-bellied marmot (*M. flaviventris*) and her young, sitting on a rock above their burrow entrance. It is shortly after sunrise, and the animals are oriented broadside to the sun, which increases radiant heat gain at this cold time of the day. (Photo by K. B. Armitage)

Fig. 3.6, *bottom*. Adult female yellow-bellied marmot (*M. flaviventris*) late in the morning, when it is considerably warmer. Her body is oriented longitudinally to the sun's rays, which minimizes radiant heat gain. (Photo by K. B. Armitage)

integrates radiation and convection to produce a value that determines whether an animal will gain or lose heat.

For larger marmots, such as *M. olympus* or *caligata*, we might predict a lower critical temperature; i.e., they should be more sensitive to high temperatures (because larger animals are more susceptible to overheating) and also better able to sustain activity at lower temperatures. Similarly, within any given species, adults might be more sensitive to overheating than yearlings, which in turn should be more sensitive than young of the year. An inverse correlation would be expected regarding susceptibility to hypothermia, especially since small individuals are highly subject to convective heat loss.

Marmots show little behavioral capacity for disposing of excess heat, although in the deserts of eastern Washington State, above-ground yellow-bellied marmots visibly salivate when temperatures exceed 32°C; the rate of dropping saliva correlates directly with ambient temperature (W. Bernds, unpublished data). Moreover, it appears that this same species is able to cool itself by sitting or lying on rocks when there is adequate convection (K. B. Armitage, personal communication). It is thus possible that young animals are able to withstand higher ambient temperatures because they are more strongly affected by convection. If so, this might correlate with the smaller body size of low-elevation species and of low-elevation populations within *M. flaviventris*.

Summary and Conclusions

1. Marmots' daily activity patterns consist of bimodal above-ground activity peaks in the morning and late afternoon during midsummer, and a unimodal activity peak at midday in the early spring, late summer, and during inclement weather; these patterns show consistent differences among age and sex classes.

2. The diel cycle also suggests that temperature constraints are important in the behavior of marmots, such that animals seek to avoid both overheating and hypothermia.

CHAPTER 4

Maintenance: Foraging, Predation, and Burrows

Although sociobiology emphasizes *social* behavior, there is every reason to suppose that evolutionary processes influence such asocial behaviors as foraging, avoiding predators, using burrows, etc. Moreover, each of these activities, when closely examined, often reveals distinctly social components: among many species, foraging is influenced by the presence of others (social facilitation), by the prospects of group detection of predators as well as cooperative defense, and so forth. Moreover, burrows are shared with others, and burrows typically are dug by many individuals. Thus, even the fundamental behaviors by which marmots maintain themselves—as opposed to their more clearly social interactions—have important implications for sociality.

Foraging

Marmots are primarily vegetarians, eating a wide range of meadow plants, although they do not forage indiscriminately. For example, when Olympic marmots emerge in late spring, they eat largely grasses, sedges (*Carex*), and occasional dug-up roots. Within a few weeks, their diet includes flowers such as lupine (*Lupinus subalpinus*), avalanche and glacier lilies (*Erythronium montanum* and *E. grandiflorum*), mountain buckwheat (*Polygonum bistortoides*), and magenta paintbrush (*Castilleja oreopola*). When meadow flowers become particularly abundant, flower heads are selectively eaten. Sapling alpine firs (*Abies lasiocarpa*) and western white pine (*Pinus monticola*) are occasionally gnawed, perhaps as much for their abrasive as their nutritive value; marmots kept in captivity without abrasive food sometimes develop pathologically overgrown incisor teeth (personal observations).

Grazing by Olympic marmots has a marked impact on subalpine vegetation, especially in the more productive wet meadows. In such habitats, marmots remain in a more confined area than in dry meadows, where they forage over a broader region (Del Moral, 1984). As forage impact increases, grass and sedge species decrease slightly in abundance and preferred species decrease markedly while unpalatable species increase greatly. The diversity of vegetation communities tends to increase with moderate grazing (Del Moral, 1984).

Hoary marmots in Alaska's Kenai Peninsula commonly eat vetches (*Oxytropis, Astragalus*), sedges, fleabane (*Erigeron*), and fescues (*Festuca*). However, foods are not simply consumed in proportion to the amount available; rather, some are clearly avoided, such as dryas (*Dryas*), willow (*Salix*), and blueberry (*Vaccinium*) (Hansen, 1975). Flowers, and frequently flower heads, notably of such species as Indian paintbrush (*Castilleja*), avalanche lily (*Erythronium grandiflorum*), western anemone (*Anemone occidentalis*), lupine (*Lupinus*), and false hellebore (*Veratrum viride*), constitute a prominent part of the diet of hoary marmots in the Washington Cascades. Grasses are predominant in the diet of Vancouver marmots, with additional large quantities of sedges and grasses in the spring and a shift to forbs, notably *Lupinus*, in the late summer; phlox (*Phlox*) is also consumed in substantial amounts (Milko, 1984). The diet of Vancouver marmots varies more between different colonies in May than later in the summer, probably because of site-specific availability (Martell and Milko, 1986); especially when food is sparse, beggars aren't choosers. As the summer progresses, Vancouver marmots—like their Olympic and hoary congeners—consume less grasses and more forbs.

Yellow-bellied marmots are similarly selective in their foraging. Captive yearlings preferentially consumed the flowers of columbine (*Aquilegia caerulea*), lupine (*Lupinus floribundus*), and larkspur (*Delphinium barbeyi*), but they avoided the shoots (Armitage, 1979), which contain a higher proportion of toxic secondary compounds, notably alkaloids. Apparently, yellow-bellied marmots (and presumably marmots in general) adjust their foraging so as to maximize input of nutrients while minimizing toxic chemicals (Freeland and Janzen, 1974).

A study of *M. flaviventris* in the White Mountains of California found that forbs are preferred over sedges and grasses, especially in the spring and early summer (Carey, 1985a). This is probably because of the former's lower cellulose content and higher amounts of protein. The fact that grasses and sedges are consumed at all (they constitute at least 25% of *M. flaviventris* diets) is probably due to their greater availability and the animals' need to maximize nutrient intake during the brief active season.

Carey (1985b) also used univariate and multivariate techniques to correlate habitat variables with the amount of time 26 yellow-bellied marmots spent foraging in 11 different areas. She found definite spatial preferences, positively correlated with food-plant biomass and negatively correlated with high, dense vegetation. This apparent contradiction suggests that the animals balance food intake with predator avoidance, since marmots foraging in high, dense vegetation are probably less able to spot potential predators.

The food preferences of woodchucks are perhaps the widest of all the marmots'. In early spring, after emergence but before vegetative shoots or flowers are available, woodchucks consume acorns, hickory nuts, and sapling buds (Meier, 1985). As the season progresses, they consume grasses and flowers, both wild and cultivated species, and virtually everything found in the traditional American garden. Woodchucks will even climb trees to feed on wild cherry (*Prunus*) blossoms (personal observation). This wide-ranging palate is characteristic of at least two Asiatic species as well: Menzbier's marmot (*M. menzbieri*) consumes 130 plant species (Maschkin, 1982; Fig. 4.1), and the long-tailed marmot eats more than 140, especially cereals, legumes, and composites (Davidov et al., 1978). (It should be emphasized, however, that documented food preferences are likely to reflect, at least in part, the area sampled as well as actual preferences of the animals.)

Marmots are also occasionally carnivorous. The stomach of one yellow-bellied marmot collected in Oregon was filled with sphinx moth caterpillars (*Deilophila lineata*) (V. Bailey, 1936). On two different occasions in the spring, I saw an Olympic marmot carrying a dead chipmunk (*Eutamias townsendi*) in its mouth. Chipmunks are abundant in most marmot colonies, and the two species appear oblivious to each other's presence; certainly, a marmot is too slow to catch a chipmunk above ground. But in the spring, while digging for roots, marmots might conceivably unearth and kill a late, shallow hibernator. Of course, they might also eat carrion. The tarbagan eats locusts when forage is scarce (Nekipelov, 1978), and, according to one report, 28% of the spring diet of *M. camtschatica* consists of (unspecified) animal food (Kapitonov, 1960).

I have identified adult marmot bones and hair in some marmot scats collected in early June, at colonies of both *M. olympus* and *M. caligata*. This apparent cannibalism is probably the result of scavenging, since I have rarely seen fights of these species result in even minor scarring or

Fig. 4.1. Adult Menzbier's marmot (*M. menzbieri*). This species, occupying a very restricted range within the USSR, has been reported to consume a particularly wide variety of plant species. (Photo by V. Maschkin)

loss of blood. Infanticide is ruled out in these cases, since the young are not born until nearly a month later; at such time, however, infanticide is a possibility, although I have no documented cases for either *M. olympus* or *M. caligata*. A simple test suggested that meat was not a preferred food item but was taken when vegetation was in short supply: I presented sliced beef hearts at each of three different *M. olympus* colonies on three different days and recorded whether the meat was taken within 2 hours by any animal in the colony. On May 19, the meat was eaten at two colonies, both of which were snow-covered, but was ignored at a third, which was essentially snow-free; on June 27, the meat was accepted at the one colony that was still snow-covered but not at either of the others, which were then snow-free; on July 14 all colonies were snow-free and the meat was not consumed at any.

Olympic and hoary marmots drink almost daily where standing water is present, and they frequently eat snow, especially just after seasonal emergence. However, water is generally lacking at many colonies, and the animals apparently obtain moisture from the meadow plants and/or the abundant morning dew. By late August and early September the

TABLE 4.1

Grass-bringing by M. olympus

| | Times grass brought | | | |
Grass bringer	June	July	August	Total
Parous female (8)	29	33	8	70
Nonparous female (6)	5	6	3	14
Adult male (6)	8	6	10	24
Two-year-old (9)	3	3	6	12
Yearling (16)	6	17	10	33
Young of the year (29)	–	–	0	0

meadows are noticeably drier than their early-season condition, and dietary desiccation may well be related to the onset of hibernation.

I observed both Olympic and hoary marmots bringing grass into their burrows. The 153 observations for *M. olympus* are summarized in Table 4.1. Before her litter's emergence the parous female was the predominant grass gatherer, but once the young of the year began foraging for themselves, her gathering frequency declined drastically and paralleled that of the other animals. Most of the grass collected by all animals was dry and brown; this is particularly striking after the third week in June when the colony meadows become quite green and gatherers make long journeys to obtain dry material. This grass is presumably used for either food or bedding, with the latter more likely (Fig. 4.2).

Home range (or territory size) and forage quality are inversely correlated among many animals, presumably because individuals are selected to restrict their activities to the space that is necessary to achieve their needs but to avoid unnecessarily large ranges, which would impose additional and unnecessary costs. Accordingly, when food reserves are available within a smaller area, cost/benefit considerations should favor restricting one's activities to that smaller area. I supplemented the forage quality at three hoary marmot colonies by depositing 70 kg/hectare of ammonium nitrate on a region of 2 hectares at each colony in mid-August of one year, after the method of Holmes (1984a). During June of the following year, adult females (both parous and nonparous), two-year-olds, and yearlings spent significantly more time within those two treated 2-hectare regions than they had the previous year, and significantly more time than comparable individuals spent in any 2-hectare region of three untreated control colonies.[1] By contrast, adult males in the treated colonies did not restrict their ranging in comparison to those in

[1] binomial test, P < .05 in each case

Fig. 4.2. Reproductive adult female Olympic marmot (*M. olympus*), gathering grass in late June. This material is then brought into the natal burrow. (Photo by D. P. Barash)

the untreated colonies, suggesting that for them, considerations other than foraging influenced their use of space (see Chapter 8). By July, the patterns of habitat use by adult males at the treated colonies were significantly different from those at the untreated colonies[2] and not different from those of other residents at the treated colonies.[3]

Woodchucks also tend to have smaller home ranges when forage quality is better (Grizzell, 1955; de Vos and Gillespie, 1960). A well-studied Pennsylvania population responded to a decrease in the natural food supply by an increase in individual home ranges and a corresponding reduction in population density. A similar pattern may well apply to other marmots. Among long-tailed marmots, for example, alpine-dwelling animals maintain substantially larger home ranges than do forest-dwellers; this has been attributed to the lower density of forage in alpine habitat (Davidov et al., 1978).

Vegetative standing crop tends to be higher at greater distance from the home burrows of Alaskan hoary marmots (Holmes, 1984a). On the

[2]binomial test, $P < .05$
[3]chi-square, $P > .10$

other hand, foraging at greater distance not only requires a greater expenditure of energy, but also subjects the animals to greater risk, since these regions tend not only to be farther from natural refugia among rocks but also to have fewer escape burrows dug by the marmots themselves. There tends to be a negative correlation between intensity of patch use and the frequency of selected plants available after 6 weeks of hoary marmot grazing (Holmes, 1984a); this is not surprising, given that arctic-alpine habitats regenerate slowly (Bliss et al., 1973).

The role of caloric density in marmot foraging strategies remains to be illuminated, but it may well be significant, as herbivores are generally limited by the rate at which they process food in the gut rather than the rate at which they ingest nutrients *per se* (Westoby, 1974). Food availability, especially during gestation, seems likely to be a limiting factor for the reproductive success of individual female marmots (Downhower and Armitage, 1971; Andersen, Armitage, and Hoffmann, 1976).

Eight years' data on *M. flaviventris* showed that body weight in June is strongly influenced by the date of last accumulated snow; i.e., when animals are able to begin foraging earlier in the spring, they weigh more (Armitage, Downhower, and Svendsen, 1976). Ten years of data on *M. caligata* reveal a similar correlation.[4] For both *M. flaviventris* and *caligata*, there is no significant correlation between June body weight and time of the first accumulated snow the previous autumn, although there is a significant correlation for both species between June weight and length of snow cover.[5] It seems clear, however, that length of snow cover is also correlated with date of last accumulated snow. For both species, when the correlation of June body weight with date of last accumulated snow is partialed out, Kendall partial rank correlations were low ($-.06$ to $-.229$); this suggests that yearly variation in June body weights of both *M. flaviventris* and *caligata* is primarily a function of when marmots begin feeding and hence gaining weight in the spring rather than duration of winter *per se* (Fig. 4.3).

Young woodchucks in New York State are weaned between early May and early July. This range occurs partly because some yearling woodchucks reproduce, and yearlings emerge from hibernation later than do adults. In a closely monitored population of *M. flaviventris* in the Rocky Mountains of Colorado, young of the year may be weaned between mid-June and early August, although weaning in the first half of July is most frequent (Armitage, Downhower, and Svendsen, 1976). Among

[4]Spearman rank test, $P < .05$ for all age and sex classes
[5]Spearman rank test, $P < .05$

Fig. 4.3. Reproductive adult female Menzbier's marmot (*M. menzbieri*) with her young of the year. Note the abundant vegetation at the time of the young's emergence. (Photo by A. Nikolskii)

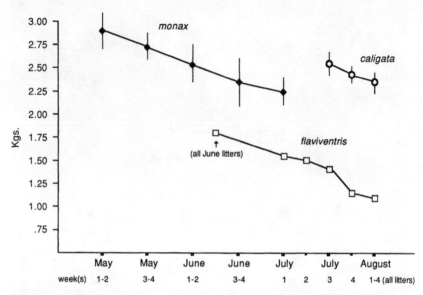

Fig. 4.4. Correlations between mid-September body weight and date of weaning for three marmot species. (Data on *M. flaviventris* from Armitage, Downhower, and Svendsen, 1976)

the hoary marmots of the Washington Cascades, the above-ground season is the most compressed of the three species: young of the year are weaned between the third week of July and the first week of August. For all three species, body weight achieved by September 15 is highly correlated with time of weaning: animals weaned late weigh significantly less, presumably because they have less foraging time available (Fig. 4.4).[6] As we shall see, weight at hibernation is significantly related to overwinter mortality (Chapter 11).

Alpine and subalpine meadows would appear to be regions of restricted primary productivity offering relatively few niche opportunities. This should be especially true of the severe talus slopes that are preferred as burrow sites by the montane marmot species. Throughout much of western North America such habitats commonly support three different medium-sized herbivores: pikas (*Ochotona princeps*), marmots, and ground squirrels (*Spermophilus* spp.) often occupy the same rockslide and live within several meters of each other. A 23,000-square-meter talus slope at 2,100 meters elevation in western Montana was occupied by

[6]Spearman rank correlation, P < .05 in each case

13 hoary marmots and at least eight pikas and 13 Columbian ground squirrels (*S. columbianus*), while a comparable region at 3,850 meters in central Colorado supported 8 yellow-bellied marmots, 7 pikas, and 10 Wyoming ground squirrels (*S. elegans*) (Barash, 1973b). Thus, with decrease in latitude, *M. flaviventris* replaced *M. caligata* and one species of *Spermophilus* replaced another, whereas pikas were found at both sites.

Ecological separation among these species was similar at the two sites. In both cases, the marmots—whatever the species—were intermediate between pikas and ground squirrels with regard to the proportion of their above-ground time spent on the talus as opposed to the surrounding meadow. At both sites, marmots concentrated their foraging between 5 and 25 meters of the talus, whereas the pikas foraged primarily within the talus itself. And at both sites, the marmots (and ground squirrels) did most of their foraging during late June and July, whereas pikas foraged most intensely during August.

Alpine and subalpine environments provide "boom and bust" situations of nutrient availability: abundant lush vegetation in the short summers and very little nourishment in winter. A distinction can be made between nutrient intake necessary for regular body maintenance and growth during summer and the added demands, also made during the summer, for sufficient fat to enable successful hibernation. Meadow vegetation is almost certainly supersufficient for summer maintenance; however, some degree of interspecies competition can be expected, especially over nourishment for winter maintenance, given that foraging for approximately 9 months must be telescoped into only 3 months or so. It is therefore interesting that the heaviest demands for forage appear to be made at different times by the three primary talus residents, with marmots and ground squirrels emphasizing the earlier, lush growth and pikas concentrating on the later, drier vegetation. Pikas appear to be the most closely coupled ecologically to their talus environment, with ground squirrels the most independent and the various marmot species generally intermediate.

Predation

Woodchucks are preyed upon most heavily by red foxes (*Vulpes fulva*) (Scott and Klimstra, 1955), domestic dogs, and human hunters. Meier (1985) observed predation by great horned owls (*Bubo virginianus*); in northern New Brunswick, woodchucks are an important part of coyote (*Canis latrans*) diet during May and June. I have also seen red-tailed hawks

(*Buteo jamaicensis*) attempt, unsuccessfully, to take woodchucks. Black bears (*Ursus americanus*) have been known to dig out occasional hibernators in late fall or early spring, and young woodchucks are said to be found on occasion in the stomachs of large rattlesnakes (*Crotalus*).

In eastern Washington State, yellow-bellied marmots make up over 70% of the total biomass consumed by nesting golden eagles (*Aquila chrysaetos*). Over a seven-year period, *M. flaviventris* constituted 355 kg out of a total of 393 kg (90%) of mammalian prey brought to golden eagle nests (Marr and Knight, 1983). Coyotes are the most notable ground predators, although Armitage (1982a) witnessed only two cases of successful predation in 20 years of observations in Wyoming and Colorado. Badgers (*Taxidea taxus*) appear to be especially formidable predators, because of their digging ability. In one case, a badger dug into an occupied yellow-bellied marmot burrow and destroyed an adult female and two litters, totaling seven young (Armitage, 1984). Yellow-bellied marmots also have been seen to chase weasels (*Mustela frenata*) and martens (*Martes americana*) (Kilgore, 1972). Although weasels are doubtless too small to prey upon adults, evidence suggests that they are occasionally effective predators on young of the year and even, in one case, on a yearling (K. B. Armitage, personal communication).

I have seen a cougar (*Felis concolor*) and a coyote each take an adult Olympic marmot. In both cases, the predator approached to within about 15 meters of the victim, working its way to nearby clumps of alpine fir; the marmots were captured after rapid downhill runs, before they could flee to their burrows. The following incident is also of interest: A coyote elicited an alarm call upon approaching an Olympic marmot colony, whereupon the marmots all ran to burrows. When the coyote rushed the nearest one, it entered its burrow, leaving the coyote scratching furiously outside. After a few minutes another marmot about 7 meters away looked out of its burrow, saw the coyote and called, whereupon the coyote ran to this new burrow, but again to no avail. This was repeated four times, after which a frustrated coyote left the area to the marmots and an amused observer. This may have been a young coyote; however, adults can be effective predators. I identified marmot hairs in two of 36 coyote scats collected.

A fisher (*Martes pennanti*) once traversed an Olympic marmot colony and evoked alarm calls. Black bears are common in the Olympic Mountains and were observed in marmot colonies on four occasions; the marmots appeared to ignore the bears unless they approached to within about 6 meters, when an alarm call generally followed.

I have seen hoary marmot predation by golden eagles, lynx (*Lynx canadensis*), and coyotes in Washington State, and have witnessed a grizzly bear (*Ursus horribilis*) attempt unsuccessfully to unearth a hoary marmot in Glacier National Park, Montana. A wolverine (*Gulo gulo*) collected on Alaska's Kenai Peninsula had hoary marmot remains in its stomach (Rausch and Rausch, 1971). Golden eagles may well be effective predators on Vancouver marmots; an average of 0.42 overflights and attacks per hour in the spring have been reported (Milko, 1984), declining to 0.18 per hour in late summer.

The eagle owl (*Bubo bubo*) is a reported predator upon *M. marmota* (Muller-Using, 1956), and snow leopards (*Felis uncia*) prey upon *M. himalayana* (Novikov, 1956) and *M. caudata* (Schaller, 1977) in the summer, when ungulates are more difficult to capture. Wolves (*Canis lupus*) are also predators on these two species (Schaller, 1977) and also on *M. sibirica*, *camtschatica*, and *bobak*, as well as on the Nearctic *M. broweri*. Golden eagles appear to be suitable predators upon the Palearctic species as well, and brown bears (*Ursus arctos*) are known to have dug holes in pursuit of *M. caudata* in the mountains of northern Pakistan (Schaller, 1977).

In the alert posture, characteristic of all marmots, individuals sit up on their haunches and survey their environment (Figs. 4.5–4.8). It seems likely that this posture facilitates spotting of predators. Although it is popularly believed that marmots use "sentries" to warn against predators, there is no evidence that particular animals serve in this capacity. Among Olympic and hoary marmots, reproductive females are the most watchful animal during the 2-week period of their young's emergence, while the adult male demonstrates consistently high levels, probably associated with male/male competition (Fig. 4.9). Resident adult yellow-bellied marmot females spend about 24% of their foraging time in the alert posture, whereas peripheral adult females—not integrated into a colony with other adults—average 38% (Svendsen, 1974). Given that peripherals engage in virtually no social interactions with other animals, their higher frequency of alert behavior suggests that the behavior is related to predator avoidance, which requires greater alertness among solitary individuals. Satellite male hoary marmots, which are not spatially associated with other colony members, show a similar pattern: they spend 31% of their foraging time in the alert posture, whereas colony adults average 20% during the same time periods.[7] Similarly, Formozov (1966) has written that in *M. bobak*, "marmots that have no near neighbors (for instance, after a poison campaign in centers of zoonosis infection) become very easily frightened and greatly shorten their grazing pe-

[7]binomial test, $P < .05$; $n = 5$ and 13; 101 observation-hours

Fig. 4.5, *upper left*. Up-alert posture by reproductive adult female woodchuck (*M. monax*). Note the unusually prominent enlarged nipple. (Photo by Pennsylvania Game Commission)

Fig. 4.6, *upper right*. Up-alert posture by Alpine marmot (*M. marmota*), in Vanoise National Park, France. (Photo by D. P. Barash)

Fig. 4.7, *lower left*. Up-alert posture by gray marmot (*M. baibacina*). (Photo by unknown Soviet biologist)

Fig. 4.8, *lower right*. Up-alert posture by black-capped marmot (*M. camtschatica*). (Photo by unknown Soviet biologist)

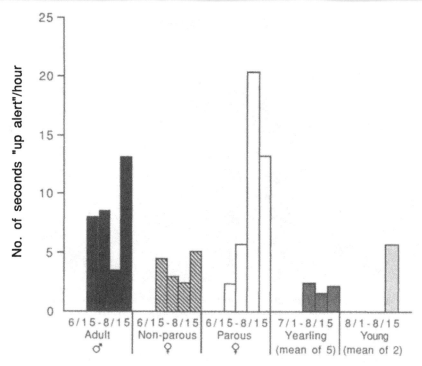

Fig. 4.9. Proportion of above-ground activity by Olympic marmots (*M. olympus*) in the up-alert posture. Data are grouped in 2-week intervals and based on a minimum of 2 individuals and 2 hours of observation during each interval.

riods. They have little chance to accumulate the amount of fat necessary for hibernation and winter survival. Such lone individuals sometimes migrate for hundreds of meters to establish, together with other wandering marmots, a new colony."

Both Olympic and hoary marmots appear to seek each other's company and feed together preferentially. Feeding groups of up to 8 animals at a time frequently occur, although these groups are loosely organized and constantly changing in composition and location. Different animals enter and leave throughout the day, and at times no grouping is discernible. Such groups are most closely aggregated early in the daily activity period when the animals are just moving into the meadows from their burrows; as feeding becomes more pronounced, particularly in the late afternoon, the associations become more diffuse. However, there is a definite tendency for the yearlings in particular to congregate while feed-

Fig. 4.10. Feeding group of three Olympic marmots (*M. olympus*). The animal on the right has its head down, eating, while the two on the left are looking up. (Photo by D. P. Barash)

ing, especially early in the season. It seems likely that such grouping provides the opportunity for additional warning about predators (Fig. 4.10).

Marmot behavior has a curiously discontinuous character, like a motion picture that is stopped about every five seconds as the animals look up, typically remaining on all fours while chewing their food. These pauses may well provide opportunities for the visual communication important in the maintenance of individual distances and social integration, as well as possibly checking for predators. Yellow-bellied marmots foraging in the periphery of their colony look up significantly more often per minute than do those in the center (Armitage, 1962). Among Olympic marmots, the amount of time spent looking up per minute is independent of the animals' proximity to particular burrows. However, proximity to other animals is important: adults in feeding groups averaged 14.8 seconds looking up per minute of feeding (SD = 3.4), whereas the same animals when solitary averaged 33.1 seconds (SD = 9.9); the difference is significant.[8] A similar correlation, with significant differences, existed among yearlings: 12.9 seconds (SD = 3.1) per minute

[8] t test, $P < .01$; based on 31 different feeding groups and 82 individuals

looking up when in feeding groups, compared to 27.6 seconds (SD = 9.0) when feeding alone.[9]

Decreased looking up when an animal is in a feeding group might be due to any combination of the following: (1) less need to look up, if looking up indicates a response to potential predators; (2) social facilitation of feeding, if looking up is regarded as merely an interruption of feeding; or (3) enhancement of communication among group members, if looking up involves looking at each other in addition to looking for possible predators. In addition, foraging itself is rarely continuous, being interrupted by running to a burrow generally about every 30 to 60 minutes. Among Olympic or hoary marmots, these interruptions are usually correlated with some external disturbances such as the sudden appearance of deer (*Odocoileus hemionus*) or the call of a raven (*Corvus corax*) or flicker (*Colaptes auratus*).

A similar correlation between nearest neighbor distance and looking up while foraging occurs for *M. caligata* as well. When yearlings were within 7 meters of another individual during July, they averaged 6.2 seconds looking up per minute (SD = 3.5), whereas when no other animal was within 7 meters, they averaged 9.0 seconds looking up per minute (SD = 4.1); these data are suggestive but not statistically significant.[10] Adult male hoary marmots during late May and the first two weeks of June averaged 13.3 seconds (SD = 6.7) looking up per minute, whereas adult females averaged only 5.0 seconds (SD = 3.8); this difference is significant.[11] By contrast, during the first two weeks of August, males and females showed no differences in this respect (males: 7.3 seconds; females: 6.9 seconds).[12] It seems likely, therefore, that males' looking up is influenced not only by predator avoidance but also by other considerations, most likely male/male competition and reproductive behavior early in the year (see Chapter 8); by late summer, both males and females are influenced primarily by predator avoidance, with probably little if any male/female differences. Thus, the change in the frequency of looking up between late May/early June and late August is not statistically significant for females, whereas it is significant for males,[13] presumably because predator risk remains relatively constant throughout the year, whereas the social benefit of alertness applies chiefly to males, and only early in

---

[9]t test, P < .05
[10]t test, P < .10; n = 38
[11]t test, P < .05; n = 13 males and 18 females
[12]n = 9 males and 11 females
[13]t test, P < .05; n = 8 males and 13 females

the year. Furthermore, whereas proximity to another individual (within 7 meters) significantly decreases looking-up behavior by adult females, its effect is insignificant, although in the same direction, for adult males. Once again, this suggests that among adult males, antipredator alertness is alloyed with social considerations to a greater degree than for adult females.

Yellow-bellied marmots in the White Mountains of California spend about 10% of their time looking up while foraging; like Olympic and hoary marmots, they look up significantly less when foraging in a social group (Carey and Moore, 1986). This study also found that yearlings and young of the year did more looking up than did adults, and that the former were generally more influenced by factors suggesting predation risk. Since juveniles are smaller than adults, they may well be more susceptible to predation, especially by smaller predators such as weasels or *Buteo* hawks.

Since woodchucks are essentially solitary, foraging groups are not generally found. In a sense, all woodchucks are "peripheral" individuals, without benefit of the additional antipredator vigilance provided by other group members, and also without the likely benefit of another's alarm calling (see Chapter 10). I recorded the time spent looking up by 12 yearling woodchucks during early May; they averaged 12.1 seconds per minute, which exceeds the time spent looking up found for either hoary or Olympic marmots of the same age at a comparable time following their emergence.[14] These findings are consistent with an interpretation that by virtue of its solitary existence, *M. monax* is more vigilant for predators than are the more social, montane species. However, conclusions may be misleading in such cross-species comparisons, because the frequency and effectiveness of predators must also be compared.

The montane species achieve safety in part by retreating to talus slopes and boulder fields, within which they are less accessible to most predators. Hoary marmots are more likely than ground squirrels or pikas to use large boulders as lookouts (Tyser, 1980), presumably because they are less susceptible than the latter species to predation because of their large size; hence they can afford to be more conspicuous. For marmots, then, it may well be that the cost/benefit ratio for seeing potential predators but at the same time making oneself more visible to them is less than that for the smaller meadow residents.

There appears to be risk associated with foraging in regions of high vegetation: yellow-bellied marmots spend more time looking up—and

[14] t tests, P < .10

not foraging—when they are in tall grass (Carey, 1983). The microhabitat choice of Vancouver marmots is not predictable from the abundances of preferred forage species because risk of predation by golden eagles appears to exert strong influence on foraging locations, inducing marmots to avoid areas that they might otherwise be expected to prefer (Milko, 1984). Milko (1984) did not address foraging and patch selection differences as a function of size and age class; as we have seen, larger animals can be expected to be at less predator risk while foraging. Adult hoary marmots, not surprisingly, forage farther from the talus than do yearlings or young of the year, and also engage in longer foraging bouts (Holmes, 1984a), although this may also be a reflection of their larger gut capacity.

Most studies of optimal foraging have concentrated on the various energetic costs and benefits but have given scant attention to possible risk factors. It seems likely, however, that the risk of being preyed upon while foraging is an important consideration for many species, especially animals such as marmots, which are eaten by a variety of abundant predators throughout their ranges. With this in mind, a study of hoary marmots in Alaska measured "forage quality" in different patches of meadow habitat by assessing (1) the frequency of all plant species, (2) the frequency of selected plants, and (3) the percentage of selected plants in a number of patches. Similarly, "risk factors" were evaluated for each patch by determining (1) the number of refuge burrows, (2) the mean distance to the nearest refuge burrow, and (3) the distance from each patch to the home talus. Patch use was found to correlate positively with forage quality and negatively with risk factors. However, food abundance and risk were also strongly intercorrelated (Holmes, 1984a).

In an effort to separate forage and risk factors, Holmes (1984a) enhanced forage quality by applying fertilizer, and as a result patch use increased sixfold. However, the marmots also manipulated the variables: they dug additional refuge burrows in patches where forage quality had been increased, thereby reducing the risk of foraging there. It seems clear that whereas foraging and predator avoidance are each important and conceptually distinct factors in the basic biology of marmots, the two cannot be considered in isolation from each other.

Burrows

Marmot colonies typically include several burrow systems, within which the animals spend about 80% of their lives. Burrows provide a place for hibernation, sleeping, and rearing young, a stable thermal environment and refuge from both predators and excessive social encounters with conspecifics. Marmots typically dig burrows in open meadows or talus slopes, often using the natural interstices among large boulders, tree roots, or buildings. Those used for residences have more than one opening. Smaller unoccupied auxiliary burrows generally are much shallower than residence burrows, typically less than 150 cm deep. In most cases these refuge burrows have only one opening and are used only temporarily, in an emergency. Winter burrows of long-tailed marmots average 2.8–3.2 meters deep, whereas summer burrows are about 0.8–1.6 meters deep; the difference presumably is due to the thermal insulation required of hibernacula (Davidov et al., 1978). One yellow-bellied marmot colony had 78 burrow openings in a 0.85 hectare area (Svendsen, 1974). Three hoary marmot colonies in Washington State had respectively 32, 45, and 49 burrow openings per hectare.

Partial excavation of an Olympic marmot home burrow revealed sleeping chambers lined with sedges, lupines, and various other meadow plants to a depth of 25 to 35 cm. Distinct latrine chambers were also found, with dung piled 10–15 cm deep and, unlike the sleeping chambers, without covering vegetation (Beltz and Booth, 1952). Such vegetation averages 5–6 kg per sleeping chamber among long-tailed marmots (Davidov et al., 1978). Separation of sleeping from latrine chambers is characteristic of marmot burrows in general. Among all species, trails often connect frequently used burrows (Fig. 4.11).

M. olympus burrows are most often located in open meadows and characteristically have a dirt "porch" on the downhill side. This porch, composed of earth and rocks removed when the burrow is dug, is generally somewhat elevated and gradually increases in size during the summer, as the burrows are enlarged and the debris is piled outside (Fig. 4.12). Adults in particular spend considerable time lying on these porches, especially in the morning and evening, before leaving and entering their burrows. The raised porches, along with large rocks, tree stumps, or other elevated objects, also serve as lookout sites.

A heavy growth of sedges frequently adorns these porches. Olympic marmots nibble only the tender tips of the sedges, with little apparent effect on the plants' growth. The combination of disturbed soil, regular fertilization, and very light grazing apparently makes this "porchside" habitat especially attractive to these plants. The characteristic growth of sedges around their burrows may be advantageous to the marmots, providing cover and shade. As the snow melts during the summer and the marmots extend their feeding areas, auxiliary burrows are dug and

Fig. 4.11. Adult Olympic marmot (*M. olympus*) at the downhill aspect of a large burrow opening. Note the distinct trail running from lower right to upper left. (Photo by D. P. Barash)

Fig. 4.12. Burrow of Olympic marmot (*M. olympus*) in early July, showing moderate development of dirt porch downhill of the opening. (Photo by D. P. Barash)

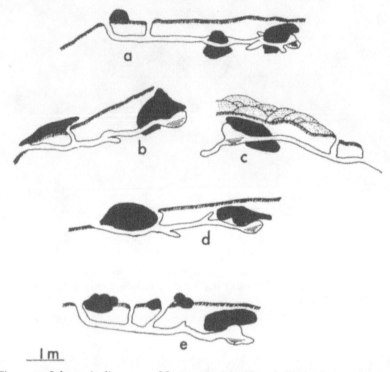

Fig. 4.13. Schematic diagrams of five excavated yellow-bellied marmot (*M. flavi-ventris*) burrows. Burrows a through d were occupied by adult females, burrow e by an adult male. Dark objects are rocks. (Modified slightly from Svendsen, 1976)

enlarged throughout the colony; as a result, the animals are always within a few seconds' run of a burrow.

By contrast with Olympic marmots, which often dig burrows in the middle of grassy meadows, hoary and yellow-bellied marmots are more likely to employ rockpiles or other large objects, digging their burrows in the interstices and thence into the ground. Olympic, hoary, and yellow-bellied marmots most commonly inhabit burrows in slopes of between 15 and 50 degrees; there is no evidence that particularly flat or steep terrain is preferred. A typical *M. flaviventris* burrow has a main entrance (up to three) extending to about 0.6 meters below ground, after which the main passageway extends about another 4 meters horizontally. The nest chamber is located at the end of the tunnel, and several short blind tunnels branch off from the main passageway (Svendsen, 1976; Fig. 4.13).

Temperatures within the burrow vary little from 10°C between June and October (Kilgore, 1972).

Woodchucks prefer well-drained sites for their burrows, and when all are equally dry, they prefer steep slopes to flat areas, presumably because of lessened danger of flooding. Woodchucks only rarely hibernate in open fields; rather, they tend to overwinter in brushy or lightly forested sites, moving to meadows and hillsides in the spring following emergence. This may be due to the longer persistence of snow at woodland sites, which provides better insulation during the winter. Or perhaps they offer better drainage. Woodchuck burrows may have from 1 to 5 entrances, and are characterized by a mound of fresh earth at the entrance and trails leading to foraging areas. Burrows typically descend steeply for about 1 meter, whereupon they level out; at a depth of 20 cm they are about 25 cm in diameter. Woodchuck burrows are often deeper than those of the North American montane species, rarely being less than 1 meter in depth and commonly reaching 2 meters or more. One burrow system was about 15 meters long (Grizzell, 1955; Fig. 4.14). Woodchuck burrows are also used by many other species, including cottontail rabbits (*Sylvilagus floridanus*), skunks (*Mephitis mephitis*), opossums (*Didelphis virginiana*), red and gray foxes (*Vulpes fulva* and *Urocyon cinereoargenteus*), and raccoons (*Procyon lotor*). Each year, about 7–8% of occupied woodchuck burrow systems are newly dug and slightly more than 50% of existing burrow systems are in active use (de Vos and Gillespie, 1960). There is no evidence that older woodchucks occupy burrows that are significantly different from those of the yearlings (Merriam, 1971), although dispersing young of the year not surprisingly dig shallow, one-entrance burrows.

The montane species appear to dig fewer new burrows per year than does *M. monax*. Soil is generally much shallower in subalpine and alpine meadows than in woodchuck habitat; the montane species may well dig shallower burrows because digging is considerably more difficult in their environment. Among the montane species in particular, burrows are used from year to year by succeeding generations, which is also consistent with likely difficulty of digging new ones. The record seems to belong to the Asiatic bobak marmot, among which some sites have been used continuously for centuries, and whose dirt "porches" are immense, extending up to 20 meters in diameter and several meters in height (Formozov, 1966). In fact, the Asiatic species seem to be extraordinary in the general magnitude of their burrow construction: hibernation burrows of *M. baibacina* and *caudata* reach depths of 5 to 7 meters (Chekalin, 1965; Kizilov and Semenova, 1967)—as opposed to the typi-

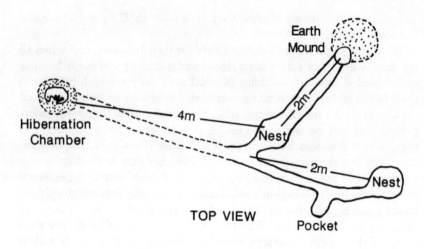

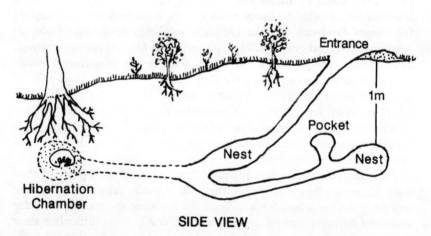

Fig. 4.14. Schematic diagram of an excavated woodchuck (*M. monax*) burrow. (Modified slightly from Grizzell, 1955)

cal 1–2 meters for *M. monax* and less than 1 meter for *M. flaviventris* and *olympus*. It is unclear whether this correlates with winter snow depths or ambient temperatures. One burrow of *M. camtschatica* was 113 meters long (Kapitonov, 1960).

  M. broweri colony residents typically hibernate in a single den (Rausch and Rausch, 1971) as do *M. camtschatica* (Kapitonov, 1960) and *olympus*.

M. caligata is more variable in this respect: colony residents used two different burrow systems in three out of nine cases in which I have been able to determine the hibernating arrangements of hoary marmots. In the others, all animals overwintered in one, although it was not possible to determine the actual "sleeping" arrangements. High-elevation yellow-bellied marmots show a tendency to aggregate during hibernation, but as in the hoary marmot, colony members do not always share the same hibernaculum (Andersen, Armitage, and Hoffmann, 1976). Woodchucks generally hibernate alone. There may be a thermal benefit in communal hibernation, but this remains to be demonstrated; alternatively, communal hibernation may simply be a response to a shortage of adequate hibernacula.

To some extent, all marmots seem to wall off their hibernacula with a plug composed of earth, feces, and vegetation, which might provide both insulation and some protection from predators; whereas these plugs are often quite rudimentary in *M. monax* burrows, they can be from 1.5 to 3 meters thick in *M. bobak* burrows, requiring 2–3 weeks for the marmots to construct (Shubin, Abelentsev, and Semikhatova, 1978). Heavy mechanized equipment is needed to dig through such structures (Formozov, 1966).

Hibernacula may also be used as the primary summer burrows, but not necessarily. When woodchuck hibernacula are far from preferred foraging areas, a short seasonal migration occurs in spring with the animals moving from their hibernating burrows to their summer residences and then back again in fall (personal observations in New York State). However, such movements did not take place in a southern Maryland population (Grizzell, 1955). Hoary and Olympic marmots typically use the hibernaculum as a major occupied burrow during the above-ground season, and hibernacula are often the largest and oldest such burrows, as suggested by the size of the dirt porch.

One colony of *M. caligata* studied in Montana was unique in that the animals changed their location during the summer (Barash, 1974a). They emerged from hibernation in the talus and rocky ledges overlooking a basin of rich subalpine meadow. This latter area received a heavier snowpack (3 meters in mid-June) than the relatively clear, windblown meadows around most other marmot colonies. As the season progressed, the animals moved down into the basin, following the melting snow into the newly exposed meadow and using previously existing burrows as they were revealed. This seasonal movement covered nearly 0.5 km and an altitude drop of 200 meters. During the movement, burrows en route were used by the same animals for periods varying from 1 to 20 days, and the

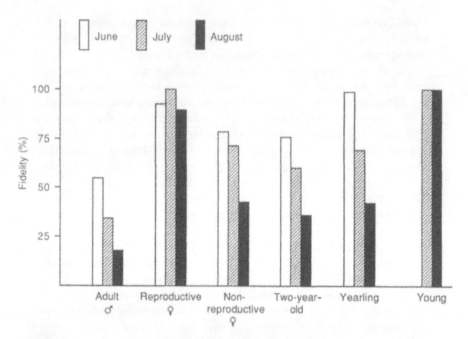

Fig. 4.15. Burrow fidelity of *M. caligata*. Calculated as the percentage of either morning or evening observations at which nightly occupancy of a particular most commonly used burrow was confirmed. Based on a minimum of 11 different observations for each of four different individuals in each cohort for each month.

same burrow was often occupied by several animals at different times. By late July a new colony arrangement was established in the lower basin.

Among both *M. olympus* and *caligata*, reproductive females, both before and after parturition, and their young of the year typically show great fidelity to a single sleeping burrow throughout the above-ground season. By contrast, other animals generally are more variable, with high burrow fidelity early in the season followed by a gradual decline (Fig. 4.15). The adult males consistently show the lowest burrow fidelity;[15] all classes except the parous females and young of the year show a significant decline in burrow fidelity from June to August.[16]

It has been suggested that hibernacula may be limiting resources for marmots (Andersen, Armitage, and Hoffmann, 1976); this notion may

[15] binomial tests, P < .05
[16] Spearman rank test, P < .05 for adult males, nonparous females, two-year-olds, and yearlings

well repay further study. It remains to be seen, for example, to what extent hibernacula are bequeathed to resident marmots by a variably beneficent habitat and to what extent they can be produced on demand by their occupants. For example, the distance between hibernacula was nearly 500 meters among monogamous Alaskan hoary marmots, leading to the suggestion that monogamy may result when inter-hibernaculum distance is too great for an adult male to defend more than one, combined with the fact that forage at a single site is generally insufficient to support more than one adult female and her offspring (Holmes, 1984b). On the other hand, if hibernacula can be constructed more or less as needed, then high inter-hibernaculum distance may simply be a result of the existing pattern of social spacing rather than a cause. It also remains to be seen whether hibernacula must meet certain physical requirements in order to be viable and, hence, whether their potential abundance can be estimated as an environmental resource independent of other aspects of each species' biology.

Summary and Conclusions

1. Marmots are primarily vegetarians, preferring certain grasses and forbs over others; occasionally they eat meat as well.

2. Forage quality and home range tend to be inversely related, and above-ground forage time correlates with pre-hibernation body weight; there is some niche separation between marmots and other alpine and subalpine inhabitants.

3. Marmots have numerous predators; antipredator vigilance includes alert postures and interruption of their foraging to look up.

4. Antipredator vigilance is reduced when other individuals are nearby, suggesting that individuals derive a possible benefit from group living.

5. Risk avoidance appears to exert a strong influence on marmot life, especially through foraging patterns.

6. Burrows are an important aspect of marmot biology, although at present, studies of marmot burrow architecture have not progressed beyond the descriptive stage.

PART

II

Social Behavior

Water is H$_2$O, hydrogen two parts, oxygen one,
But there is also a third thing, that makes it water
And nobody knows what that is.
                    D. H. Lawrence, *The Third Thing*

Bees are not as busy as we think they are. They
    just can't buzz any slower.
                    Kin Hubbard, *Abe Martin's Sayings*

The organization of individuals in time and space reflects their social structure and, in turn, influences it as well. To some extent, variability in marmot social structure is the rule rather than the exception: living arrangements change from year to year within the same population, and they also differ from one population to another, even in the same year. Certain patterns even change unpredictably within a given population, whereas others vary predictably as the yearly cycle progresses. On balance, and despite the numerous idiosyncratic vagaries, it is still possible to characterize most marmot species in a meaningful way, one that also identifies significant biological distinctions between species. To a great extent, as we shall see, these distinctions correlate with the environments occupied, and in which the various marmot species have presumably evolved. These correlations, moreover, seem to be meaningful in contributing to fitness; that is, they appear to be *adaptive*.

CHAPTER 5

Social Structure

Woodchucks are the most solitary of all marmots. By contrast, all of the Eurasian species appear to be at least moderately social, and many, like the Alpine marmot (*M. marmota*) and apparently the black-capped marmot (*M. camtschatica*), are highly social. Of the six North American species, and, quite possibly, of all the world's species, *M. monax* is the only one that does not live in readily identifiable social units or colonies. Among woodchucks, consistent social structure seems limited to the very brief association between adult male and adult female during the early spring breeding season and to the mother/young nexus, which terminates at weaning when young of the year typically disperse. At least in Ohio, however, some adult males maintain territories that include one or more adult females (Meier, 1985).

During three years' observations of woodchucks, I noticed yearlings foraging together (within about 5 meters) on four occasions, an adult female and a yearling twice, and an adult male and a yearling once. During the course of extensive studies of free-living woodchucks in southern Pennsylvania, two animals were noted at the same burrow in 23 cases: 16 of these were adult male/adult female pairings during early spring (F. H. Bronson, 1964). Adult male/adult female associations are always brief, following the female's emergence in mid-March. I noted 14 instances of adult males and adult females sharing the same burrow. The median duration of such associations was 4 days; the longest was 11 days. In 12 of the 14 cases, the male moved in with the female, with the co-residence occurring in the female's hibernation burrow; one occurred in the male's hibernation burrow; and one occurred in a previously unoccupied burrow. Following these consortships, the males departed in all

14 cases. In five of these, the males established sequential consortships, each with one other female (see Chapter 6). Impregnated females remained more sedentary than the adult males, which generally continued to travel, presumably in search of additional females. Eleven different adult males occupied an average of 3.3 different burrows each during a 3-week period in late March and early April; 12 different adult females occupied an average of 1.2 different burrows each during the same time.

The adult male woodchucks studied with radio telemetry by Meier (1985) in Ohio can be classified as either "residents" or "wanderers." The former remain relatively sedentary, with a loose social structure including one or two females, occupying home ranges that averaged 10,440 square meters (SD = 2,633). They visited the hibernacula within their home ranges on a daily basis; such hibernacula averaged 104 meters apart (SD = 25). These males engaged in primarily affiliative interactions with the adult females and juveniles encountered within their home ranges. By contrast, wandering adult males didn't interact with resident animals; they used burrows that were employed only infrequently by other individuals. A similar pattern has been suggested for the black-capped marmot: nonbreeding adult males wander widely, spending no more than 1–3 nights in a burrow (Kapitonov, 1978).

During the period between spring emergence and May 1, adult female woodchucks maintained home ranges that were considerably smaller than the males': 2,512 square meters (SD = 761). Later in the summer, however, they expanded their ranges to 13,485 square meters (SD = 2,408), larger than the adult males'. The relatively large early-season home ranges of resident male woodchucks may be due to their searching for and defending adult females, whereas the subsequent expansion of adult females' home ranges is likely due to their need for adequate forage during and after lactation.

In southern Ontario, woodchuck density in suitable habitat averages about 0.5 animals per hectare (de Vos and Gillespie, 1960). In central New York, I calculated a density of approximately 1 animal per hectare. By contrast, the population density among yellow-bellied marmots in occupied habitat at 2,900 meters in Colorado is 5.8 animals per hectare (Kilgore and Armitage, 1978), and among hoary and Olympic marmots in Washington state, occupied meadows average 7.8 and 9.2 animals per hectare, respectively.

Despite the woodchuck's solitary nature and generally low population density, even this species shows a rudimentary social tendency. Thus, a study of 112 occupied burrows in New York State revealed a contagious distribution, with an indistinct but nonetheless nonrandom clumping of burrows (Merriam, 1971). Movements among neighboring burrows were more frequent than between them, which suggested to Merriam that woodchucks might be exhibiting a form of vestigial coloniality and the operation of positive, attractive social factors in addition to simple avoidance and outright aggression.

Captive woodchucks maintained in visual contact with each other gained significantly more weight than did isolates, suggesting a stimulative effect of social grouping, even in this least social of marmots (E. D. Bailey, 1965b). Foraging is known to be a "contagious" behavior among many animals, and marmots—even woodchucks—appear to be no exception. Captive woodchucks in visual contact with each other were also more likely to synchronize their activities than were individuals who were visually isolated. This tendency declined significantly from April until November; i.e., the difference between "visual contacts" and "visual isolates" diminished as the season progressed; individuals became more self-involved and less other-involved as hibernation approached (E. D. Bailey, 1965b).

I trapped 28 adult female woodchucks in central New York State, exposed them to each other in groups of two, and recorded their behavior as a function of the distance between the individuals when trapped. Each individual was tested only once. Of the seven tested pairs that were originally trapped less than 600 meters apart from each other, only one pair actually fought; the other six each showed clear dominance/subordinance relationships, as revealed by threat (raised tail, erected hair, occasional hissing and tooth-chattering) on the part of the dominant, along with prompt avoidance by the subordinate. Of the seven pairs of woodchucks trapped more than 1,000 meters apart from each other, four pairs fought, two showed mutual threat, and only one showed clear dominance/subordinance. These findings are analogous to others obtained for raccoons (*Procyon lotor*) and red foxes (*Vulpes fulva*) (Barash, 1975a), and suggest that among woodchucks, as among these so-called "solitary" carnivores, some degree of local familiarity occurs, such that social relationships are apparently established among neighbors even though interactions among free-living adults may be infrequent.

Yellow-bellied marmots, in contrast to woodchucks, typically live as colony members: one adult territorial male, one or more adult females, their young of the year, and yearlings, which generally disperse early in the summer (Armitage, 1962, 1974, 1975; Downhower and Armitage, 1971; Armitage and Downhower, 1974). There may be substantial variability in the number of colony residents, not only between colonies, but in the same colony from year to year (Fig. 5.1), depending on the number of adult females present, whether they are reproductive, the number of young and yearlings produced, and whether dispersal of the latter has been complete.

Peripheral individuals also exist, including single or paired individuals, or sometimes a female and her young; these animals remain isolated from colony residents, interacting only for occasional matings, if that (Svendsen, 1974). It seems likely, in fact, that a continuum of social structure exists in *M. flaviventris*, ranging from solitary individuals to one-harem colonies to continuous multi-harem sites (Johns and Armitage, 1979). Even in the latter cases, however, individuals are not homogeneously distributed: a given marmot can readily be identified as belonging to one harem or another. Finally, transient individuals are also found. Presumably these are dispersers that have not yet established them-

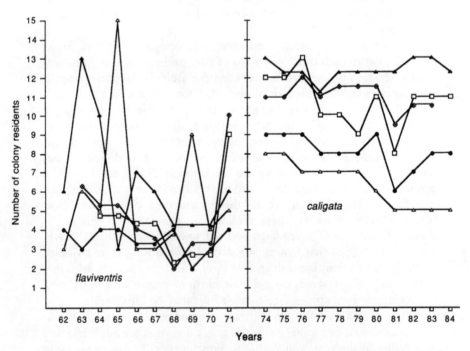

Fig. 5.1. Annual variation in colony size among five different *M. flaviventris* and *M. caligata* colonies. (Data for *flaviventris* from Armitage, 1975)

Fig. 5.2. Grouping of four Olympic marmots (*M. olympus*); from the left, they are an adult male, two adult females, and a yearling. (Photo by D. P. Barash)

selves in a residence and, occasionally, former colony residents who have departed during adulthood. In a well-studied site in the Colorado Rockies, 75% of all yellow-bellied marmots were colony residents, 16% were peripherals, 8% were transients, and 1% were uncategorized (Svendsen, 1974).

The basic social structure of Olympic and hoary marmots is indistinguishable, consisting of one adult male (occasionally a second, "satellite" male), one or more adult females, two-year-olds, yearlings, and young of the year (Fig. 5.2). Females reproduce biennially—every other year—and yearlings remain in their natal colony, dispersing the following year as two-year-olds. Each female is therefore typically associated with either young or yearlings (occasionally neither), but never both. Reproductive females occupy their own burrow, eventually sharing it with their young. Nonreproductive females may reside with the adult male, with yearlings, or with two-year-olds. Yearlings typically reside together, but they may also reside with two-year-olds or, rarely, with the adult male (Barash, 1973a). A similar population and age structure appears to exist among European Alpine marmots as well: of 4 colonies, 1 was monogamous, 3 were bigamous, and of the latter, 1 contained an additional, satellite male (Barash, 1976b). By contrast with *M. flaviventris*, individual colonies of *M. caligata* tend to be relatively consistent in their year-to-year population size (Fig. 5.1); the variability of *M. caligata* colonies is significantly less than that of *M. flaviventris* colonies.[1]

In addition to the typical isolated colony, both Olympic and hoary marmots are also found in "colony town" situations, in which numerous colonies are contiguous to each other, and interactions occur daily between residents of different colonies. Most colony town residents can readily be identified as members of a particular constituent colony; more than 95% of social interactions occur within a given colony (Holmes, 1984b). Colony structure itself may shift between monogamy and polygamy, depending upon annual recruitment and mortality of adult females. Among Olympic marmots, I observed 38 different colony-years (one colony-year equals one colony observed for one year) distributed as 16 different colonies, of which 9 were observed for 3 years each. There were 7 monogamous colonies, 22 bigamous colonies, and 1 trigamous colony; the remaining 8 were composed of either solitary females with offspring (3) or were groups lacking adults (5). In south-central Alaska, 11 *M. caligata* colonies were all found to be monogamous groups (Holmes, 1984b). In Washington State, of 74 *M. caligata* colony-years, 24 were monogamous, 46 bigamous, 1 trigamous, and 3 undetermined.

For both the Washington State Olympic and hoary marmots, there were no tendencies for either monogamy or bigamy to be more stable; i.e., the relative proportions of each social structure remained comparable from year to year,[2] although there was some switching, both from monogamy to bigamy and vice versa. Bigamous colonies become monogamous when an adult female dies; monogamous colonies become bigamous when a two-year-old female does not disperse and remains in her natal colony, becoming an adult, three-year-old female.

A second, "satellite" adult male occurred in 7 of the 38 Olympic marmot colony-years and in 10 of the 74 hoary marmot colony-years. Satellite males were socially subordinate to the colony male during the first half of each summer, occupying a burrow at the colony's periphery. As hibernation approached and the breeding season receded, satellites were increasingly incorporated into the colony's social structure (see Chapter 9). Hoary marmot satellite males were younger (3.2 years, SD = 0.4) than colony males (4.8 years, SD = 0.5)[3] and were more likely to be found at larger harems than at smaller ones: only 1 satellite was found at the 24 monogamous colonies, as opposed to 8 satellites at the 46 bigamous ones.

[1] F-max test, $P < .05$
[2] Kruskal-Wallis analysis of variance, $P > .10$
[3] Mann-Whitney U test, $P < .05$; 8 satellite and 13 colony males

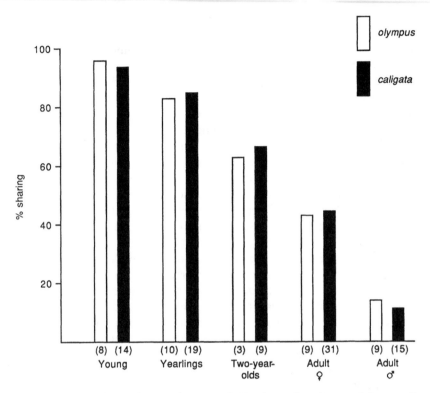

Fig. 5.3. Burrow sharing by *M. olympus* and *M. caligata*. Based on a minimum of 51 observations for each cohort, at which individuals were identified as either occupying their burrows alone or sharing with others.

Like most semifossorial animals, marmots are relatively sedentary, and they typically associate with an identified burrow, burrow system, and foraging meadow. Among both Olympic and hoary marmots, older individuals are less likely to share their sleeping burrows (Fig. 5.3).[4] Adult males share their burrows significantly less than does any other age or sex class.[5] Prior to July 1, no satellite male was ever seen to share his burrow with another marmot.

The frequency of transients apparently varies among the different species. During ten years' observation of hoary marmots in Washington State, I observed a total of just 17 transient individuals, approximately 2% of the observed resident population. By contrast, transients among

[4] Spearman rank test, P < .05 for both species
[5] binomial test, P < .05

yellow-bellied marmots constitute about 8% of the population (Svendsen, 1974). I have never observed a resident adult hoary or Olympic marmot leave its colony. About 15% of my sampled woodchuck population consisted of transients at any given time. I assume that these are equivalent to Meier's (1985) "wandering" individuals. He reported 17 resident adult males and 5 wanderers in an Ohio population. It appears that in the progression from *caligata* to *flaviventris* to *monax*, there is a steady increase in the ratio of transient to resident individuals.

This seems reasonable because a high proportion of female woodchucks breed annually, whereas the proportion of breeding females is lower among yellow-bellied marmots and lower yet among hoary marmots. Virtually every adult female woodchuck therefore produces a set of transient, dispersing offspring every year. Some of the observed wanderers may thus be dispersers; others may be adults who retain a transient life style.

Yellow-bellied marmots, by contrast, often skip a reproductive year, and hoary and Olympic marmots regularly do so; sometimes they skip consecutive years as well. Furthermore, since young woodchucks disperse during their first year, prior to hibernation, and since (as we shall see) overwinter mortality is roughly comparable for all marmot species, the resident woodchuck population regularly experiences an additionally high level of transient dispersers. For yellow-bellied marmots, by contrast, a year of potential overwinter mortality intervenes before dispersal occurs. For hoary and Olympic marmots, would-be dispersers must pass through two successful hibernations. Although most transient marmots are dispersing juveniles, adult woodchucks occasionally disperse, leaving their home burrow, apparently never to return; adult yellow-bellied marmots rarely do this, and adult Olympic and hoary marmots are not known to do so at all.

Not surprisingly, interactions between hoary marmot residents and transients make up less than 0.01% of their social interactions. For woodchucks, I observed a total of 18%. This frequency does not differ significantly from the proportion of transients observed in the population,[6] whereas for *M. caligata* the difference is significant;[7] this suggests that woodchucks in their asociality do not discriminate between transients and residents, at least with regard to frequency of encounters, whereas hoary marmots do. Data on this point are unavailable for yellow-bellied

[6] binomial test, P > .05
[7] binomial test, P < .01

marmots, but we can predict them to be intermediate between *M. monax* and the other two species.

The selective pressures giving rise to the various marmot social systems will be examined throughout this book, and a more general scenario will be proposed in the concluding section. Regardless of the actual phylogenetic sequence, or the selective pressures involved, the contrast between yellow-bellied marmot and woodchuck social structure can at least be visualized by imagining a population of woodchucks that then experiences increased social attraction—or, at least, tolerance—among adult females, while the adult males strive to retain as many females as possible within their territories. In addition, if the young delay their maturation and dispersal, thereby remaining with their mothers during their first year, and if all this takes place in a montane or steppe habitat, *M. monax* social structure will have been converted into that of *M. flaviventris*. Note that the actual evolutionary sequence may have been in the other direction, with *M. monax* derived from a *flaviventris*-like system, by decreasing social tolerance and speeding up development, or indeed, the two may conceivably have been independently evolved from a different ancestral pattern.

A good picture of Olympic and hoary marmot social structures, in turn—if not necessarily a picture of the actual phylogenetic sequence—can be gained by continuing the hypothetical trend from *M. monax* to *M. flaviventris*: increasing the social attraction among all individuals, except the adult males, and further delaying maturation and dispersal so that colonies consist of young, yearlings, and two-year-olds, in addition to at least one adult male and one or more adult females, all of whom are more closely integrated both spatially and socially than yellow-bellied marmots.

Summary and Conclusions

1. Woodchucks are basically solitary, but with hints of an occasional social tendency.

2. Yellow-bellied marmots are loosely colonial, their social structure being developed around a male with his harem of one or more females, their offspring, and dispersing yearlings.

3. Olympic and hoary marmots are more densely colonial, social systems being monogamous or bigamous (rarely trigamous), and consisting of a male with his harem, young of the year, yearlings, and dispersing two-year-olds.

4. Satellite males and colony towns are found among both Olympic and hoary marmots, and peripheral individuals are found among yellow-bellied marmots.

5. Woodchucks experience the highest frequency of transients, followed by yellow-bellied marmots and finally Olympic and hoary marmots, which experience the lowest such frequency. This may be a cause and/or an effect of social intolerance.

Descriptive Anatomy of Social Behavior

T he study of animal behavior is in many ways more difficult than the study of physics, at least partly because the basic units under study are less discrete. Animals are more complex than billiard balls or quarks, and their behavior, at first blush, is frustratingly difficult to describe or even to identify. But after a while, certain patterns begin to emerge—at least to the human observer; we assume that these patterns are of significance, even if not of conscious significance, to the animals as well. Only after the basic anatomy of marmot social behavior is identified can we begin to recognize "meta-patterns," patterns in the patterns of behavior. Then, with grounding in the basics of marmot social behavior, it becomes possible to identify interspecies patterns in behavior within the genus *Marmota*, patterns which themselves might eventually begin to make sense as part of another, higher-order meta-pattern, the adaptive significance of animal social behavior in general.

Greeting

The most conspicuous social interaction among high-elevation marmots is "greeting," a behavior that occurs among many sciurid rodents. For the highly social Olympic and hoary marmots, greeting is especially frequent. Although its exact function is unknown, greeting presumably involves olfactory and possibly tactile communication, and it seems to be especially relevant to individual recognition. Greeting occurs most often when two animals meet after a brief separation, and it is particularly frequent immediately after the animals emerge from their burrows, in the morning and then again in the afternoon.

Fig. 6.1, *top*. Nose-to-nose greeting among young hoary marmots (*M. caligata*). (Photo by D. P. Barash)

Fig. 6.2, *bottom*. Nose-to-cheek greeting among Olympic marmots (*M. olympus*). (Photo by D. P. Barash)

Fig. 6.3. Neck-chewing greeting among adult Vancouver marmots (*M. vancouverensis*). (Photo by G. W. Smith)

Among those species in which greeting is especially well developed, greeting generally begins as nose-to-nose or nose-to-mouth contact, initiator to recipient, and frequently concludes at this point (Fig. 6.1). Greetings of longer duration become progressively more complex and intense, involving successive stages of nose-to-cheek contact (Fig. 6.2), chewing on the ear, and, if the greeting continues long enough, chewing the neck of the recipient (Fig. 6.3). This behavior is not rigidly stereotyped; it varies in form and sequence. Interlocking of the teeth, occasionally producing audible clicks, often follows nose-to-nose greetings and seems to be associated with "high intensity"; i.e., greetings that involve evident agitation on the part of the animals involved and that may be associated with indications of threat or aggression.

Immediately preceding and during a greeting, the animals generally retract the split upper lip and expose the upper incisors. This behavior also occurs when an animal "sniffs at" another from a distance, or when the animals are nervous or disturbed, as by a predator or sudden sound or movement by the observer. Most commonly, greeting animals stand on all fours. Yearlings frequently assume a distinctly stiff-legged posture while engaged with larger animals. However, variability is the rule and an animal occasionally gives or receives a greeting with little sign of agitation, by simply turning its head, sometimes lethargically, as while lying down.

The Olympic marmot seems to prefer initiating to receiving a greeting. Greetings are often begun with great alacrity by any member of the colony, whereas the recipient usually appears much less enthusiastic. In such cases, the recipient either ignores the initiator and walks right past him or her, stands rigidly still, returns the greeting, or even briefly attacks the greeter. The recipient is more likely to initiate such an attack the longer the greeting takes, i.e., the farther it proceeds. Thus, during a June sampling period, the frequency of greetings followed by fights increased progressively with increasing complexity of the greeting at termination: 11% of 467 nose-to-nose greetings were followed by attacks, as compared to 27% of 133 nose-to-cheek greetings, 61% of 18 ear-chewing greetings, and 100% of 9 neck-chewing greetings.

Actually, very few of the above "attacks" are worthy of the name. They almost always consist of a brief nip or an open-mouthed growl; therefore, the term "reprimand" would probably be more appropriate. The extent of the greeting seems to determine whether or not a reprimand occurs. An Olympic marmot receiving a lengthy greeting often gradually raises its tail, and when it is pointing up, reprimands the initiator. But whereas reprimands following a greeting were preceded by tail-raising 71% of the time, fights not involving a greeting were preceded by tail-raising 20% of the time, and 11% of chases not involving a greeting were preceded by tail-raising. The relationship of tail-raising to greeting, and to aggression, is thus complex and not clearly understood.

The changing rates of greeting reflect changes in sociality throughout the day and during the summer as well. Differences between age and sex classes are also detectable (Fig. 6.4); the adult males greet most often during June and July, while yearlings' levels are conspicuously low in June, immediately following their seasonal emergence. By August, adult male greeting rates have become quite low, exceeded by those of adult females and yearlings, while the newly emerged young of the year have become the most frequent greeters.

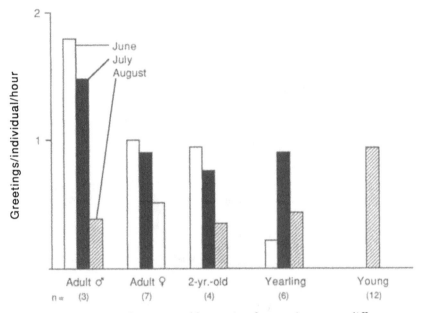

Fig. 6.4. Comparison of mean monthly greeting frequencies among different age classes of Olympic marmots (*M. olympus*). Based on a minimum of 55 animal-hours in each category.

Greeting rates for all animals combined drop significantly from June and July to August (Fig. 6.5), paralleling the decline in social activity also indicated by decreasing levels of burrow visiting, play-fighting, and social behavior. As the animals begin accumulating fat for the winter and the breeding season recedes, foraging occupies a larger part of their time, while rates of social interaction decline.

A parallel seasonal decline in the intensity of greetings, while difficult to quantify, is as distinctive and probably as important as changes in frequency: June greetings are very vigorous, active affairs, whereas August greetings are typically brief, almost casual. A random sample of 50 adult male/yearling greetings in June averaged 4.8 seconds, whereas 41 greetings after August 15 at the same time of day and among the same individuals averaged 2.9 seconds.[1] Twenty-two of these June greetings elicited reprimands or other signs of apparent agonism; by contrast, only five of the late-August greetings did so. In June 23% (12 of 52) adult male/female greetings went beyond the nose-to-nose stage, compared

[1] t test, P < .01

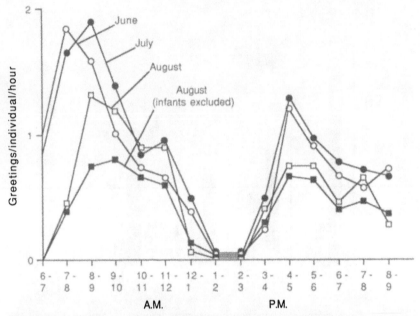

Fig. 6.5. Comparison of greeting frequency of Olympic marmots (*M. olympus*) during June, July, and August. Based on 434 observation-hours.

with only 11% (10 of 90) in August.[2] In addition, June greetings are considerably more animated than those later in the season.

Figure 6.5 illustrates another important feature of marmot social behavior, the "visiting period" that precedes morning and afternoon feeding activity. Visiting periods are especially pronounced in the early morning when the animals first emerge from their burrows. This is a time of vigorous social interaction as the animals run from burrow to burrow, greeting or being greeted by the occupant and any animal encountered en route. Patterns of greeting initiation appear to be egalitarian, not related to dominance status or home-burrow proximity. A random sample of 88 greetings occurring within 3 meters of a burrow between the burrow occupant and a newly arrived visitor showed that the occupant initiated the greetings 46 times, and the visitor, 42 times.[3]

Since only two or three burrows in a colony are normally occupied at night, most are empty during the early morning stages of visiting. The "visitor" thus becomes the temporary occupant, often waiting just out-

[2] binomial test, P < .10
[3] binomial test, P > .05

side and greeting with the next visitor who happens by, either as initiator or recipient. The resident male is particularly active at this time, entering nearly every burrow and "making the rounds" to greet or be greeted by virtually every colony member, typically many times. Thus, during 453 observation-hours of visiting periods between 0600 and 0900 hrs, 7 different resident males entered 5.2 burrows per hour (SD = 2.2) whereas 13 adult females entered 3.5 (SD = 1.6).[4] During 10 observation-days, each of three different adult males greeted with 88 to 100% of the other colony residents (mean = 94%); by contrast, 6 adult females greeted with 52 to 94% of the other residents (mean = 70%).

Following this period, feeding becomes paramount while greeting and other social behavior decline. Greeting increases slightly again in the late morning, as the animals aggregate once more just outside their burrows before the early afternoon activity lull. Most animals remain underground from about 1200 to 1500 hrs, with only an occasional aboveground straggler. A second, briefer visiting period of less intense greeting precedes the late-afternoon activity stage. Although there is some tendency to spend the afternoon lull in their sleeping burrows, about one-fourth of observed Olympic marmots regularly used other burrows during this time; consequently there was greater dispersion at afternoon emergence than in the early morning. This may contribute to the lower afternoon greeting frequencies. Greeting frequency increased again at dusk when intense feeding was over and the animals were again closely aggregated before retiring for the night.

It is generally easy to distinguish initiator from recipient during a greeting: the former actively sniffs, and moves his or her head, while the latter remains quite still, often rigidly so. I sampled Olympic marmot greetings at three different colonies from June to August, in each case distinguishing initiator and recipient (Fig. 6.6) and adjusting for the numbers of each age and sex class available in each case. The adult male was the initiator significantly more often in June than in July or August;[5] young of the year (infants) were initiators much more often than animals of any other age class,[6] and over the entire season, yearlings were initiators significantly more often than either adult males or adult females.[7]

Adult males' initiation frequency in June is almost certainly related to the breeding season, since effective sexual behavior doubtless requires evaluation of individual olfactory cues. Furthermore, wandering and dis-

[4] t test, P < .01
[5] binomial tests, P < .01 in each case
[6] binomial test, P < .01
[7] binomial tests, P < .01 in each case

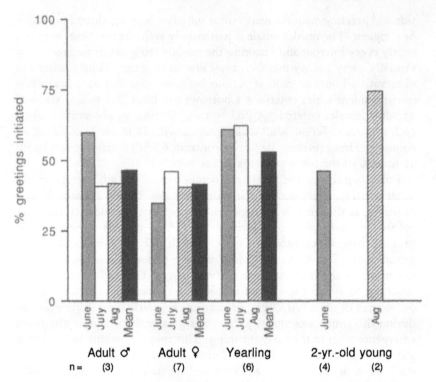

Fig. 6.6. Percentage of Olympic marmot (*M. olympus*) greetings in which individuals of each sex and age class were the initiators. Based on data from three
different colonies and no fewer than 33 greetings for each dyad. (Modified slightly
from Barash, 1973a)

persing animals are encountered early in the season: of the 8 nonresident
Olympic marmots that I observed during 3 years' intensive study, 7 were
seen during June, and of the 17 nonresident hoary marmots that I observed during 11 years, 12 were seen during June. It may well be advantageous for the resident male in particular to recognize any intruders. The
young of the year are especially active greeters, and they appear to
"bother" the other animals by initiating many sloppy and extensive
greetings (Figs. 6.7, 6.8, and 6.9).

Greeting levels and initiating frequencies do not necessarily coincide:
an animal may be involved in many greetings while being the active initiator of relatively few. Two animals may approach each other without
obvious behavioral asymmetry and either one may become the initiator,
or one (often the adult male) may approach the other from a substantial

Fig. 6.7, *top left*. Young Olympic marmot (*M. olympus*) initiating a greeting with its mother. (Photo by D. P. Barash)

Fig. 6.8, *top right*. Young yellow-bellied marmot (*M. flaviventris*) initiating a greeting with its mother. (Photo courtesy of the U.S. National Park Service)

Fig. 6.9, *bottom*. Young Alpine marmot (*M. marmota*) initiating a greeting with its mother. (Photo by D. P. Barash)

distance—clearly being the one "responsible" for the greeting—only to have the other animal step forward and become the initiator when the two are within a meter of each other. Total greeting frequency is therefore probably a better indicator of the general level of social activity, while initiation frequency reflects the specific manner in which a given animal responds to social situations in which it is involved.

Greeting seems more usefully described as an interaction than as attempted manipulation of one animal by another. It may well exemplify "probing," in which individuals extract information as to the other's identity and intentions (Owings and Hennessy, 1984). This would help explain the apparently rapid role reversals that occur: adult males frequently approach another animal, only to have that individual initiate the greeting with the adult male. In such cases, the adult male may have ascertained whether the approached animal is a colony resident or not, by its response to being approached. Then the colony member "insists" on greeting, thereby remaining familiar.

Greeting behavior may conceivably provide appeasement if initiated by young of the year or the solicitation of appeasement if initiated by the resident male. The ultimate effect of this system would thus be suppression of aggressiveness within the colony, enabling close association among potentially aggressive individuals. In fact, the extensive development of greeting behavior may be responsible for the suppression of the agonistic aspects of dominance so characteristic of the montane marmot species, with the ritualization of aggression making injurious attack unlikely and coinciding with a basically "playful" and closely integrated social system (Fig. 6.10).

Among hoary marmots, the structural aspects of greeting and the daily and seasonal cycles, as well as the relative involvement of individuals of different age and sex classes, follow a pattern similar to that of *M. olympus* (Figs. 6.11 and 6.12). Two-year-olds are involved in the fewest greetings, reflecting their reduced social integration, whereas adult males (especially in June) and young of the year are involved in the most greetings.[8] The presence of nondispersing two-year-olds results in a somewhat enlarged colony, but one composed of particularly asocial individuals. Hence, colonies with two or more two-year-olds had a lower rate of greetings per animal present per observation-hour than did colonies not containing two-year-olds (0.21 vs. 0.36), although the total number of greetings observed per colony correlates positively with increasing number of colony members.[9]

[8] analysis of variance, $P < .01$
[9] Spearman rank test, $P < .05$

Fig. 6.10. Adult female Olympic marmot (*M. olympus*) vigorously grooming the neck of one of her young, which had been persistently greeting her. A non-reproductive adult female is on the right. (Photo by D. P. Barash)

Young of the year are more likely to receive greetings from and to initiate greetings with other young than would be expected from their proportion in the population; the same applies to yearlings and two-year-olds.[10] Adult males, for their part, are equally likely to initiate greetings with adult females, two-year-olds, and yearlings,[11] but they are less likely to initiate greetings with young of the year than would be expected based on chance alone, given the relative numbers of individuals in each age class.[12] In general, adult male hoary marmots—unlike Olympic marmots—are twice as likely to initiate greetings as to receive them. Finally, adult females are more likely to initiate greetings with yearlings than would be expected from chance alone, and are less likely to do so with other adult females;[13] by contrast, their initiation rates with adult males,

[10] chi-square, P < .01; n = 35 young, 24 yearlings, and 16 adults
[11] chi-square, P > .10
[12] chi-square tests, P < .01
[13] chi-square tests, P < .01

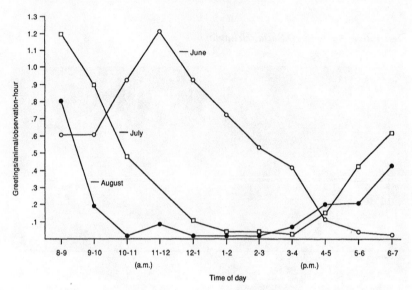

Fig. 6.11. Daily and seasonal variations in greeting frequency among hoary marmots (*M. caligata*) in Glacier National Park, Montana. Based on the following observation-hours: June, 130; July, 141; August, 73; and a minimum of 4 hours during each hour-interval of each month.

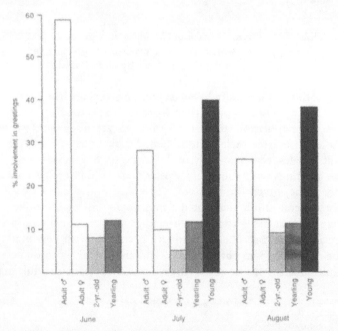

Fig. 6.12. Percentage involvement in greetings by hoary marmots (*M. caligata*). Based on the observation regime for Fig. 6.11, including four colonies altogether.

Fig. 6.13. Two yellow-bellied marmots (*M. flaviventris*) greeting each other without direct physical contact. (Photo by K. B. Armitage)

two-year-olds, and young do not differ from the expectations of chance.[14] A study of *M. caligata* in Alaska, however, revealed that adult females are significantly more likely to receive greetings than to initiate them (Holmes, 1984b).

Behavior designated as greeting occurs among yellow-bellied marmots as well (Armitage, 1962, 1965; Barash, 1973d). However, it typically does not involve mouth contact; indeed, greeting among *M. flaviventris* generally involves no contact whatever, thereby differing substantially from the greetings of both *M. olympus* and *caligata*. Rather, greeting yellow-bellied marmots commonly approach each other with tails arched and gingerly sniff at each other's faces while frequently remaining several meters apart (Fig. 6.13). Compared with their higher-elevation congeners, a lower frequency of greetings per individual characterizes yellow-bellied marmot society, with frequencies ranging from one-half to less than one-tenth of those found in *M. olympus* and *caligata*. The relative involvement of age and sex classes, however, seems comparable for all three species: highest frequencies for adult males, yearlings, and young of the year. Among the adults, higher frequencies are found early in the season. Diel

[14] chi-square tests, P > .01

cycles in greeting frequency, however, are less distinct in *M. flaviventris*, although as we have seen, daily activity cycles are comparable.

Yellow-bellied marmots appear to lack a well-developed visiting period, which suggests their lower level of social tolerance and integration. Furthermore, interaction rates among yellow-bellied marmots apparently do not vary with changes in colony population size (Armitage, 1973, 1975, 1977), although such rates are strongly influenced by individual behavioral phenotypes, with submissive individuals taking part in a high proportion of social interactions (Armitage, 1982b).

During three years' field observations of woodchucks in New York State, I observed 22 greetings: 13 were between newly emerged young, 6 between the mother and her young, and 3 between an adult male and an adult female. An equivalent number of greetings can typically be observed during a single day at an Olympic or hoary marmot colony, although admittedly the difference in observability between woodchucks and the highly social forms may also be important. During four years' field study of woodchucks in Ohio, Meier (1985) observed 69 greetings, primarily between adult males and adult females, adult males and yearling females, adult females and yearling females, adult females and adult females, and among young of the year. Adult males were never seen to greet other adult males, adult females were never seen to greet other adult females, and yearling males were never seen to greet anyone. A study of paired encounters between captive woodchucks, however, found that greeting ("nuzzling") was frequent, occurring in 68% of all encounters, and that it was initiated by the subordinate individual 63% of the time (F. H. Bronson, 1964). Under field conditions, it appears that *M. monax* typically avoids social interactions, including greeting.

"Amicable" behavior has been identified among yellow-bellied marmots, and is defined as the sum of greeting and grooming behaviors (Armitage, 1974). Certainly, the decrease in greeting frequency from *olympus/caligata* to *flaviventris* to *monax* parallels a decrease in social tolerance and, therefore, in "amicability." In *M. flaviventris*, for example, a greeting is almost never followed by a chase or a fight; it may precede grooming, play, or sexual approach by an adult male. Greetings are also virtually absent between members of different matrilines (see Chapter 10). Accordingly, it seems reasonable to interpret greetings as indicating cohesiveness or amicable (nonaggressive) relationships.

Because of the anthropomorphism of the term, however, as well as the apparent heterogeneity of greeting itself, some reservations are in order. Greetings may have a large agonistic component, as shown by the positive correlation between duration of greeting and the probability of a reprimand. Moreover, greetings seem to connote very different motivations depending on the individual involved: when done by a juvenile, a greeting appears to be toward the playful and affiliative end of a likely behavioral continuum, whereas when done by an adult it is more likely to be agonistic. In addition to age-related differences, males and females also appear to employ greetings for different purposes, also varying with the season. Among young of the year, greeting is more prominent as a play initiator for males (which are not normally considered especially "amicable") than for females (Nowicki and Armitage, 1979).

Rates of aggression, measured as the total of fights, chases, and avoidances, correlate positively with rates of "amicable" exchanges (Downhower and Armitage, 1981). However, such data do not consider the possible confounding effects of sex differences or of kinship. Following the replacement of colony males by new males, agonistic behavior, as measured by the frequency of chases, fights, and avoidances, increases, from 2.8 to 20.6 per animal per hour. At the same time, the frequency of amicable behavior—greetings plus mutual groomings—also increases, from 4.5 to 44.3 per animal per hour (Armitage, 1974).

Greeting nonetheless clearly seems to indicate something about marmot social interaction, serving as a relatively convenient and presumably meaningful indicator of social interaction. There are additional anomalies, however. Whereas European Alpine marmots evince the same basic trends as do the North American species, such as higher frequencies for adult males, reduced frequencies as the season progresses, and high frequencies for newly emerged young of the year (Barash, 1976a), overall greeting rates per animal are lower than in *M. olympus* and *caligata*, and are roughly comparable to those of the clearly less "amicable" *M. flaviventris*. Within a given species, higher frequencies of greetings appear to correlate with greater intensity in social interaction, regardless of the motivational state of the individuals concerned. But it may be misleading to assess motivation from the frequency of greetings, despite the temptation to employ it as a measure of affiliative behavior. Thus, a lower frequency of greetings correlates with a lower level of social agitation and interaction, not necessarily a lower level of social tolerance: yellow-bellied marmots that have lived together for one or more years often have few greetings, even though they use the same burrow system and often sit side by side on the same rock (K. B. Armitage, personal communication). Familiarity may or may not breed contempt, but it seems to generate a reduced greeting frequency.

Play-fights

For the observer of adult Olympic and hoary marmots, the distinction between playing and fighting is tenuous and subjective. Just as greeting gives the impression of being primarily affiliative but becomes increasingly aggressive when it becomes more lengthy and when done by older animals, "play-fighting" appears prosocial, even "amicable," when done by younger animals, but increasingly aggressive with greater duration and with greater age of the play-fighters. Like greeting, playing and fighting appear to represent a continuum, with the playing shifting to fighting among older animals.

Behaviors that seem clearly play are common among young and yearling *M. olympus* and *caligata*. They are variable, are not immediately purposive, vary dramatically in specific pattern and context, and are easily interrupted by more "serious" situations such as the appearance of a predator or an adult. Young of the year as well as yearling Olympic and hoary marmots play for extended periods (Fig. 6.14), particularly before and after intense feeding. Typically, the animals rise on their hind legs and appear to box with one another, or strain as though attempting to push the partner over backward. In this position, they often throw their heads backward and look straight up at the sky, in what may constitute a stereotyped display posture (Fig. 6.15). Not uncommonly, such upright play-fights are interrupted by a greeting, with both animals remaining on their hind legs (Fig. 6.16).

One individual may suddenly lunge at another, often leading to mock fights, especially among young of the year and yearlings, both partners prancing tense and stiff-legged toward each other, then perhaps breaking into greetings or separating suddenly, apparently in mid-sequence. Once, I saw four Olympic marmots—two adults and two yearlings—roll somersaulting together in the meadow. Adults and two-year-olds often employ the upright posture during play-fights, but young of the year do not seem to engage in upright play until 7 days after their emergence. Before this time, they play consistently on all four feet. Adults play-fight most frequently with other large individuals and never with young of the year. This may well be due to the simple physical mismatch between adults and young, which makes upright play-fighting awkward, regardless of whatever social inhibitions might be operating.

Play-fighting among adults often begins as apparent play but then becomes increasingly "serious," with growling and heightened intensity, until it appears to have all the characteristics of a fight, and often con-

Fig. 6.14, *top*. Upright play-fighting among young Olympic marmots (*M. olympus*). (Photo by D. P. Barash)

Fig. 6.15, *bottom*. Two adult female Olympic marmots (*M. olympus*) with their heads thrown back during an upright play-fight. (Photo by D. P. Barash)

cludes with one animal—generally the smaller—running away and into a burrow. Once again, these sequences blur the distinction between play and fighting.

The frequencies of upright play-fighting vary seasonally and with age and sex class (Fig. 6.17). The daily, seasonal, and age/sex related patterns of play-fighting are similar to the patterns for greeting behavior: adult males are the most active play-fighters early in the season, after which play-fighting declines steadily as the season progresses. Yearlings' play-fighting frequencies begin relatively low, then increase, before eventually declining, as do those of all other classes. Young of the year are distinctly the most active play-fighters, exceeding even the early-season levels of the adult male. Yearlings are most likely to initiate play-fights with other yearlings, and young of the year are most likely to do so with other young of the year. In particular, the latter's tendency to remain together near the burrow (see Chapter 7) may contribute to their high play-fighting frequencies, and to their high greeting frequencies as well. The decline in play-fighting among young of the year by the third week after emergence parallels their increased wandering by this time.

Fig. 6.16. Pause for a greeting during an upright play-fight between two two-year-old Olympic marmots (*M. olympus*). (Photo by D. P. Barash)

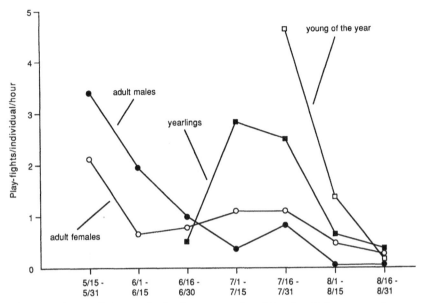

Fig. 6.17. Seasonal changes in the frequency of upright play-fights among hoary marmots (*M. caligata*). Based on a mean of 73 observation-hours during each 2-week interval and a minimum of 41 hours for each interval and eight individuals for each class.

Adult male *M. caligata* initiate more play-fights than they receive, and adult females receive more than they initiate.[15] Two-year-olds, yearlings, and young of the year receive and initiate play-fights in roughly equal frequencies (Fig. 6.18). Comparable data have been obtained for *M. caligata* in Alaska (Holmes, 1979), with the exception that young of the year engaged in fewer play-fights than did yearlings. It remains to be seen whether this difference is a function of sample size, natural variation among colonies, or some underlying difference between the Alaska and Washington populations.

The duration of play-fights also varies as a function of the age and sex of participants: play-fights involving adult males and infants are briefer than those involving adult females, two-year-olds, or yearlings (Fig. 6.19).[16] There are no significant trends relating play-fight duration to the age/sex class of the initiators and recipients.

Not surprisingly, play-fights among yellow-bellied marmots appear

[15] Mann-Whitney U tests, P < .01
[16] Mann-Whitney U test, P < .01

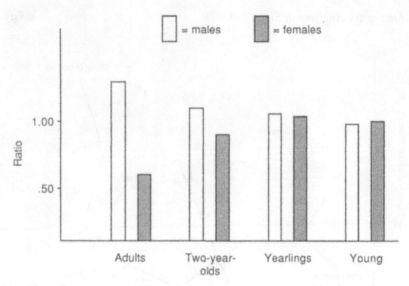

Fig. 6.18. Ratios of play-fights initiated to received among hoary marmots (*M. caligata*) in Washington State. Based on a minimum of 345 observation-hours and 12 individuals for each cohort.

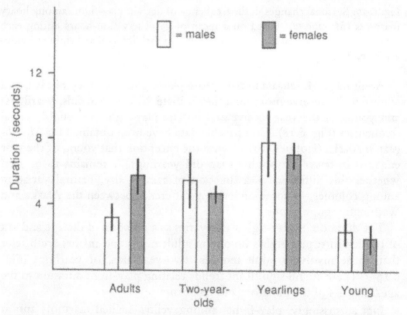

Fig. 6.19. The average duration of play-fights for hoary marmots (*M. caligata*) in Washington State. Based on the same observation regime as Fig. 6.18.

to be considerably less playful. They have been classed as agonistic behavior (Armitage, 1974, 1975, 1977) and are most prominent among
yearlings, especially males. Eighty-six percent of observed *M. flaviventris*
play-fights involved at least one male yearling, and among these yearling
duos, male/male and male/female combinations were each three times
more frequent than were female/female combinations. Homosexual play
among yearling *M. flaviventris* is significantly more frequent than heterosexual play, and males initiate play more than do females; moreover, male
play patterns involve a substantial agonistic component, and it appears
that male play is "too rough" for the females, who often seek to break off
heterosexual encounters (Jamieson and Armitage, 1987). This leads to
the conclusion that play serves especially in the enhancement of motor
skills.

By contrast, *M. caligata* play-fights among yearlings do not show any
sexually oriented pattern, consistent with the delayed maturation and
sexual development of *M. caligata* as well as with the greater agonistic
component of *M. flaviventris* behavior. Nonetheless, play-fighting among
yearling *M. flaviventris* yearlings still retains a playful character (Armitage, 1974): "The two may roll and tumble in a bout of wrestling. There
is much flailing of legs in the air and the animal on the bottom pushes
against the animal on top. . . . A moment later, their positions may be
reversed. There seems to be some nipping, but no serious biting, and evidence of inflicted pain is rare. One animal may suddenly dart away, only
to return a moment later to pounce on the animal from which it ran."

Such behavior is rare among adult yellow-bellied marmots, which do
not normally run ("playfully") toward an individual from whom they
have just run away.

I witnessed only two occasions of upright play-fighting among woodchucks; both involved young of the year that were within a few days of
dispersing. The upright posture of play-fighting involves sustained physical contact between individuals, and, accordingly, it may require a degree
of social tolerance that is poorly developed in *M. flaviventris* and virtually
unknown in *M. monax*. It is possible to conclude, alternatively, that upright play-fighting is part of the woodchuck's behavioral repertoire but
rarely shown because of physical spacing and/or low social tolerance.

Agonistic Behavior

Avoidance of one individual by another can be considered a low-level
form of agonistic behavior, and avoidance seems to be important in mar-

mot social interactions. Regrettably, however, it is often difficult to detect avoidance with certainty. Younger individuals often tend to move
aside when older animals approach, a pattern that seems especially characteristic of woodchucks; of 36 instances in which I identified adult or
yearling woodchucks walking or running toward young animals, the
latter avoided contact in 34 (94%). During one year's observations of
yellow-bellied marmots, I witnessed 18 such approaches of young animals by older individuals; of these, the former avoided the latter 12 times
(67%). During 179 hours of observation of hoary marmots during one
July, there were 57 such encounters, with the young animal avoiding the
older 22 times (38%). Chasing and fighting are both very rare among
woodchucks (Meier, 1985) and were observed between adult females
only. Taken by itself, such evidence could suggest a high level of tolerance among woodchucks. More likely, however, it indicates intolerance,
and a high level of avoidance. Avoidance appears to be a frequent pattern
among older animals, with the same trend: *monax* avoids more than *flaviventris*, which avoids more than *olympus* or *caligata*, although comparative data are not yet available.

Chases are almost certainly agonistic and are easily identified. They
may be vigorous affairs, criss-crossing the meadows at great speeds and
lasting several minutes. Alternatively, a chase may take only a few seconds, with one individual chasing another into a burrow, or it may ter-
minate after just a few meters, with both individuals remaining above
ground the whole time. Not surprisingly, chases are relatively rare among
hoary marmots, averaging about 0.01 per animal per observation-hour in
Washington State. It appears quite important, however, as a reflection
of dominance and especially male/male competition (see Chapter 8).
Among Alaska hoary marmots, about 60% of all chases were between
individuals from different colonies, despite the fact that more than 95%
of all other interactions occurred among individuals within their own
colonies (Holmes, 1984b).

As a general pattern among all marmot species, it appears that agonism declines in the late summer, with the onset of hibernation. Avoidance and chase frequencies are substantially reduced, virtually to zero in
all species for which reports exist. Among the little-known Altai marmots (*M. baibacina*), for example, "peaceful interactions among animals
from different families" occur by August (Bibikov and Berendaev, 1978).

Communication

Vocalizations

Auditory communication is especially well developed among marmots, particularly the montane forms. The best known and most characteristic sound is a "whistle," which is not really a whistle at all but rather a high-pitched tone produced by the vocal cords. The vocalizations of each species appear to be elaborated around a distinctive fundamental frequency, with different calls varying in amplitude, duration, and intercall interval. Among Olympic marmots, for example, the long or "alarm" call is a basically clear tone with an average duration of 0.39 second (SD = 0.03) and no harmonics (Fig. 6.20). The intercall interval is variable but generally exceeds 5 seconds; a single call is most commonly given.

The long call is correlated with a disturbance—either a sudden, loud

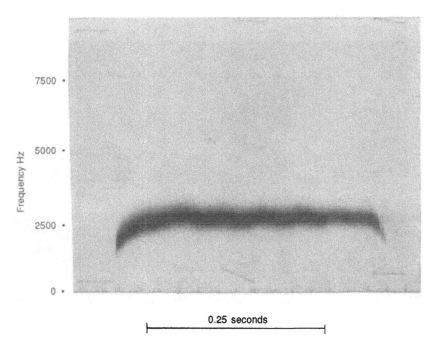

Fig. 6.20. Sonagram of an Olympic marmot (*M. olympus*) long or "alarm" call, given by a two-year-old at the entrance to his burrow.

noise or the intrusion of something new, such as a predator, into the colony (Fig. 6.21). Upon hearing this call most marmots run to a burrow; they may then enter it, or enter partially and then look outside, with their bodies either partially within the burrow or near the entrance. Animals within a burrow often look out when this call is given.

The short or "distress" call is at the other extreme of the vocalization continuum (Fig. 6.22). I have heard it given by animals taken by predators and repeatedly by trapped animals. Its average duration is 0.095 second (SD = 0.009). This call appears in a series, with variable intercall interval (mean = 0.33 second) depending upon the intensity of the stimulus. More intense stimulation generally results in a decreased interval— 0.15 second was the shortest such interval recorded. In most cases, there are at least four harmonics. When this call is given, marmots generally run to a burrow and enter, rarely pausing to look out.

A medium call is intermediate between the long and short calls, with a duration of about 0.2 second, and a variable interval, generally between 1 and 3 seconds. This call is often repeated for a prolonged period, fre-

Fig. 6.21. Olympic marmot (*M. olympus*) yearling giving a long call in response to the arrival of a Columbian black-tailed deer (*Odocoileus hemionus*). Nonpredators as well as predators often evoke such calls, especially if the intruder approaches closely. (Photo by D. P. Barash)

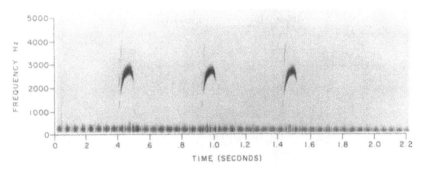

Fig. 6.22. Short call given by an adult female Olympic marmot (*M. olympus*) inside a live-trap.

quently 10–15 minutes and on one occasion nearly an hour. This call sometimes results from an earlier disruption; in other cases, no external stimulus may be apparent.

A similar array of high-pitched vocalizations occurs among the other species. Hoary marmots employ a long call at a frequency somewhat higher than that of Olympic marmots (3,400 Hz as opposed to 2,500 Hz), with a duration of 0.74 second (SD = 0.12), one harmonic, and an inter-call interval of about 15 seconds (Taulman, 1977). Short-duration, short-interval calls in response to more intense stimulation also occur. The primary vocalization of Brower's marmot is also similar, with a duration of about 0.4 second and a frequency of 4,000 Hz; in addition, a much longer "attenuated call" nearly 0.9 second long also occurs in this species (Hoffmann, Koeppl, and Nadler, 1979). The Vancouver marmot also has both long and short calls, with a fundamental frequency of 3,000 Hz. Particularly notable in this species is the elaboration of long calls as terrestrial-predator warning, and short calls as aerial-predator warnings (Heard, 1977). This is the only known instance of such specificity among marmot vocalizations.

Among the Palearctic species, *M. camtschatica* averages a long call of only 0.2 second, at 3,000 Hz. A similar fundamental pattern—a basic high-pitched call at a characteristic frequency plus more rapid pulses depending largely on the intensity of stimulation—has also been reported for the *M. bobak* "group," including *bobak*, *baibacina*, and *sibirica* (Nikolskii, 1974; Figs. 6.23 and 6.24).

High-pitched vocalizations have also been identified in detail for the yellow-bellied marmot (Waring, 1966), around a fundamental frequency of 4,000 Hz. By contrast with those of *M. olympus*, *caligata*, and *van-*

Fig. 6.23, *left*. Young bobak marmot (*M. bobak*) giving alarm call on all four feet. (Photo by N. Formozov)

Fig. 6.24, *right*. Adult bobak marmot (*M. bobak*) giving alarm call from up-alert posture. (Photo by N. Formozov)

couverensis, however, the vocalizations of *M. flaviventris* are very brief, with an average duration of only 0.037 second (SD = 0.006). As with other marmots, under conditions of high stress both the call and the intercall intervals may decrease until the animal is vocalizing as often as five times per second (Waring, 1966). Woodchucks are the least vocal of marmots, although the French Canadian name for them, *siffleur* (whistler), suggests that their vocal habits have long been noted. Woodchuck vocalizations occur between 2,700 and 4,800 Hz and are typically followed by a rapid warble. The fundamental frequency is often slurred as well (Lloyd, 1972).

Similar high-pitched vocalizations have also been reported for the Alpine marmot, *M. marmota*, and it has been claimed that these calls serve as acoustic territory markers (Bopp, 1955; Munch, 1958). However, other workers have emphasized the anti-predator function of these calls (Muller-Using, 1955; Barash, 1976a). Alpine marmots are more likely to use a repetitive series of calls when potential predators such as people or red foxes are approaching from a distance, and to give single calls when predation is imminent (Hofer and Ingold, 1984).

Marmots clearly vocalize in situations involving agitation and danger, and it seems likely that a territorial marking function, if any, is probably incidental. If alarm calling is primarily territorial, it should be most frequent early in the summer when territories are established and dispersers and other trespassing individuals are most frequent. The actual situation, however, is opposite: among Olympic marmots, for example, the number of calling sequences per hour was 0.31 in June, 0.20 in July, and 0.64 in August (Barash, 1973a). This late-season peak of *olympus* vocalization was correlated with the frequency of alarming situations—in this case, the appearance of numerous hawks, feeding on late-summer grasshopper irruptions. A similar pattern has been reported for *M. flaviventris* as well (Armitage, 1962).

The apparent alarm function of marmot vocalizations is not universal, however. Thus, during 994 hours observing woodchucks over the course of three years, I never heard a vocalization given in response to potential predators, although such predators (hawks, dogs, hunters) were relatively frequent. On the other hand, I did hear woodchucks give brief high-pitched squeals and warbles while fighting and after being released from live-traps. Given the possible role of kin selection in marmot vocalizations (see Chapter 10), it is not surprising that woodchucks emit alarm calls the least frequently of all the marmots, and that this behavior is essentially limited to selfish patterns of interpersonal communication, with little or no warning function. (On the other hand, it is also possible that woodchucks engage in very little alarm calling simply because their spacing pattern makes warning calls of little value.)

An aggressive component of marmot vocalization can also be recognized among the montane species. Yellow-bellied, hoary, and Olympic marmots occasionally "chirp" at each other, sometimes for several minutes, although more often briefly before a chase. Trapped transient male yellow-bellied marmots may sometimes be harassed by territorial males, in response to which the former chirp loudly and rapidly (K. B. Armitage, personal communication).

Low guttural growling and tooth chattering are characteristic of vigorously play-fighting animals among all the marmots. The chatter is produced by rapid movement of the lower jaw, bringing the anterior and posterior faces of the lower incisors into contact with the opposing surfaces of the upper incisors. The sound carries through all frequencies and is less than 1/100 second in duration, with an interval of about 1/10 second.

Vancouver marmots also employ another prominent vocal pattern: a call sounding like "kee-aw," usually given in a prolonged series of 100 or so, with each call separated by about 4 seconds (Heard, 1977). It apparently indicates a low-level alarm or uneasiness and is often given after alarm calling has subsided. Upon hearing this call, Vancouver marmots may run to their burrows, or they may seemingly ignore it. Olympic and hoary marmots occasionally precede alarm calling with one or two "kee-aws," but the sound—when it appears at all—rapidly grades into the more typical high-pitched whistle. Only *M. vancouverensis* employs the "kee-aw" call without its high-pitched follow-on.

Olfactory Communication

Olfactory as well as tactile information is doubtless transferred in greeting. Cheek glands, located in the oral angle, are found among all marmots (Figs. 6.25 and 6.26). Adult males and females may occasionally rub the sides of their faces on rocks or prominent vegetation, apparently as a form of scent marking. This behavior is generally correlated with high social excitement and often occurs immediately after a play-fight. Among Olympic marmots, 27 instances of cheek-rubbing by adult males were recorded in June, when social interactions were most intense, 11 were recorded in July, and only 3 were recorded in August. Similar use of cheek glands has been reported for *M. broweri* (Rausch and Rausch, 1971), *marmota* (Koenig, 1957; Munch, 1958), *baibacina* (Bibikov and Berendaev, 1978), *monax* (Meier, 1985), and *flaviventris* (Armitage, 1974, 1976).

I rubbed the cheek glands of captive hoary marmots with cotton swabs, then applied the swabs to rocks near the burrows of resident adult males. Swabbings from three different adult males elicited substantial sniffing and interest from three different resident males during June. By contrast,

Fig. 6.25, *left*. Close-up of side of the head of bobak marmot (*M. bobak*), showing cheek gland. (Photo by N. Formozov)

Fig. 6.26, *right*. Close-up of side of the head of gray marmot (*M. baibacina*), showing cheek gland. (Photo by unknown Soviet biologist)

swabbings from each of two different satellite males were virtually ignored when presented to "their" dominant male. Swabbings from strange
yearlings and young elicited very little interest, as indicated both by duration of sniffing and degree of agitation, and swabbings from yearlings
and young resident within the colony were also virtually ignored. Swabbings from strange adult females evoked interest, but there was no apparent difference in the response to reproductive and nonreproductive
females.

Cheek-rubbing may be done by either males or females, and appears
to be an indication of dominance and/or social excitement rather than
territoriality. Once again, however, the Vancouver marmot appears to be
unusual in its mode of communication: adult males are the most prominent cheek-rubbers, and they conspicuously use this behavior to mark
the boundaries of their territories, independent of such immediate social
context as social dominance or excitation (Heard, 1977).

Anal glands are also present among marmots, and the papillae are noticeably everted in some trapped, intensely aroused animals. The secre-
tion of this gland has been found to inhibit the activity of woodchucks
(Haslett, 1973); it may serve as an alarm pheromone for that species and,
conceivably, other marmots as well. This gland might be employed in
genital or anal sniffing and in the conspicuous "tail raising" behavior
(discussed below). In addition, frequent "up-alert" postures, found in
all marmot species, may provide ample opportunities for anal-gland
scent marking. A pungent odor typically characterizes marmot burrows,
and among woodchucks, intensification of this odor often precedes the
daily above-ground appearance of the animals. Among the *Spermophilus*
ground squirrels, there is a correlation between the frequency of scent
marking and the degree of sociality (Kivett, Murie, and Steiner, 1976);
our knowledge of chemical communication among marmots is at present
too sparse to indicate whether such a correlation holds for this genus
as well.

Visual Signals

The clearest visual signal among marmots is the upraised tail. Marmot
tails are large and bushy, and when raised above the body, they are conspicuous. An upraised tail appears to signal aggressiveness and may often
precede a brief lunge at another individual. In addition, marmots often
flick their tails up and down immediately before taking a few steps. Marmots commonly move by discontinuous sequences of steps, and each
sequence may or may not be preceded by a tail flip. Among Olympic

marmots, the occurrence of tail-flipping is significantly correlated with the age and sex of the animal and whether it is feeding or moving, according to the progression adult males > adult females > yearlings > young of the year. Feeding animals, of any age and sex, are less likely to tail-flip than are moving animals, further suggesting that tail-flipping serves a social function, possibly associated with dominance or, at least, attention getting. Among yellow-bellied marmots, adult males tail-flip conspicuously while patrolling their territorial boundaries (Armitage, 1974); more conspicuous yet is a characteristic side-to-side tail flagging, especially prominent in adult males (K. B. Armitage, personal communication). The arched and upraised tail is a visual signal indicating aggressive arousal in both the hoary and Vancouver marmots as well (Heard, 1977).

Marmots employ a rich variety of communication signals. They are suitable subjects for comparative analyses of signaling systems, with perhaps particular attention to possible correlations between semiotic and social complexity.

Mating Behavior

Marmot mating behavior is not easily observed. Among most species, mating occurs within a few weeks after emergence from hibernation, when observation conditions are difficult. Many if not most functional matings may take place within the burrows. Nonetheless, there is enough above-ground activity to yield some generalizations. The basic components of mating behavior are quite simple, involving male approaches to the female, female approaches to the male, male sniffing the female (often chewing on her shoulders and back), attempted mounts with the male grasping the female dorsoventrally with his forelegs about her middle while holding her back and shoulder fur in his mouth, female mounts of the male, male chasing female, male/female play-fight, and apparently successful mounts. The latter may last from 30 seconds to 8 minutes, during which the mounted animal responds by assuming a lordotic posture and arching its tail while holding it distinctly to one side (Figs. 6.27, 6.28 and 6.29).

These behaviors, although linked, do not follow in predictable sequences; rather, they can best be described by means of a probabilistic flow-chart which also incorporates changes in the frequency of transitions as the mating season proceeds. Figure 6.30 shows a flow-chart of the mating behavior of three reproductive Olympic marmots, and Fig.

Figs. 6.27–6.29. Mating behavior in Olympic marmots (*M. olympus*). 6.27, *top*: The male (facing camera, behind female) had greeted the female and chewed her ear and shoulder area, and is now chewing on her haunches; note the orientation of the female's tail. 6.28, *middle*: The male, on the left, is about to mount the female. 6.29, *bottom*: The male is now mounting the female. (Photos by D. P. Barash)

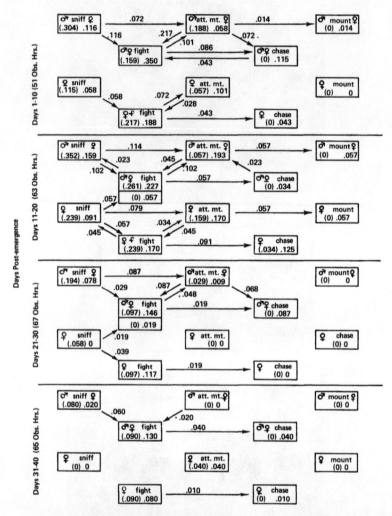

Fig. 6.30. Flow chart of sexual behavior by reproductive female Olympic marmots (*M. olympus*); see text for details.

6.31 presents a comparable chart for a nonreproductive female. The flow-charts may be entered anywhere, since marmot mating behavior does not follow a unitary course or begin consistently at any one point. Parentheses indicate behaviors beginning at a given point, while the accompanying number represents behaviors ending there. Incoming and outgoing arrows indicate transitions. Thus, the sum of the parenthetic number

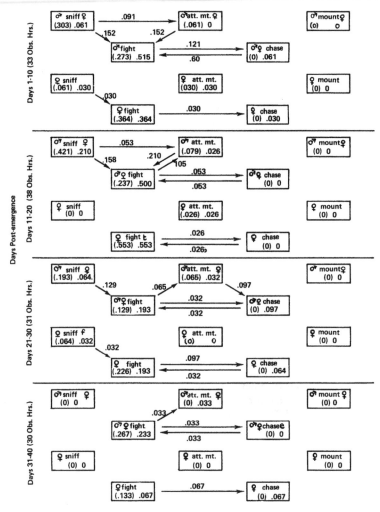

Fig. 6.31. Flow chart of sexual behavior by a nonreproductive female Olympic marmot (*M. olympus*); see text for details.

and all incoming arrows (or the sum of the free number and all outgoing arrows) equals the total number of the given behaviors observed. The upper 5 categories in each 10-day period represent behaviors initiated by the resident male toward the female concerned, while the lower 5 represent behaviors initiated by the female toward any colony member.

Similar data were obtained for six reproductive and eight nonreproduc-

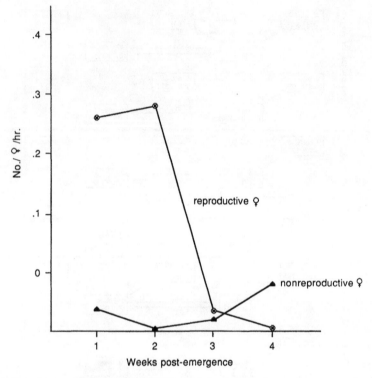

Fig. 6.32. Mounting of other individuals by reproductive and nonreproductive adult female hoary marmots (*M. caligata*). Based on six different reproductive females and eight different nonreproductive females, and a minimum of 46 observation-hours during each week post-emergence.

tive hoary marmots, with results comparable to that found for Olympic marmots: compared to nonreproductive females, reproductive female hoary marmots are more likely to mount other animals (Fig. 6.32) and to sniff the genitals of other animals (Fig. 6.33). Surprisingly, there is no difference between males' attempted mounts of reproductive and nonreproductive females, although the two types of females differ in their response to male-initiated sexual attempts: nearly all attempted mounts of nonreproductive females by adult males evoke a fight, whereas reproductive females are considerably more tolerant of male sexual behavior, particularly during the first two weeks after emergence (Fig. 6.34). The marked increase in aggressive responses by reproductive females following the second week after emergence reflects the increased aggressiveness of pregnant females. The percentage of genital sniffing by males

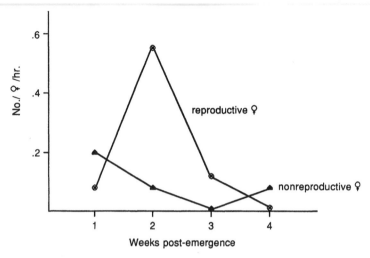

Fig. 6.33. Sniffing the genitals of other individuals by reproductive and non-reproductive female hoary marmots (*M. caligata*). Based on the same observation regime as Fig. 6.32.

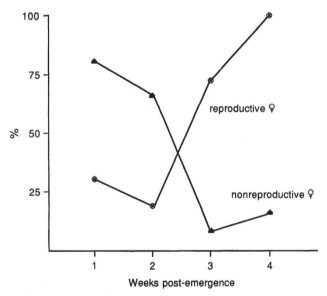

Fig. 6.34. The percentage of mounts by adult male hoary marmots (*M. caligata*) that are followed by fights between reproductive or nonreproductive adult females and the adult male. Based on the same observation regime as Fig. 6.32.

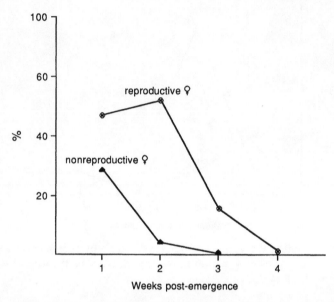

Fig. 6.35. The percentage of genital sniffs of adult female hoary marmots (*M. caligata*) by adult males that led to attempted mounts. Based on the same observation regime as Fig. 6.32.

which leads to attempted mounts indicates male evaluation of female sexual state: again, there is a substantial difference between reproductive and nonreproductive females in this respect (Fig. 6.35). These data all suggest an estrous cycle in the reproductive females with a behavioral peak (presumably coinciding with ovulation) about 1–2 weeks after emergence. During 689 observation-hours of hoary marmot behavior between mid-May and the end of June, I noted 24 male/female copulation attempts, of which 9 appeared to be successful; all but two of the attempts and all the successes involved reproductive females and all of these occurred during the first two weeks. I only observed six mounts by adult male Olympic marmots, all involving reproductive females and five of these during days 11–20.

It is clear that sniffing, fighting, and chasing are closely interwoven in marmot sexual behavior. The decrease in transitions among Olympic marmots by days 31–40 (indicated by the decline in the number of arrows in Figs. 6.30 and 6.31) bespeaks a decline in the interrelatedness of these behaviors. With the passing of the mating season, sexual behaviors become more sporadic and independent.

when adult sexual activity has generally ceased. Yearling Olympic marmots mounted adult males 12 times, two-year-olds 5 times, other yearlings 11 times and young of the year twice. Young attempted to mount other young twice, and their mother 4 times. In such cases, there is no apparent difference (either in behavior patterns or frequency) between males and females, either as initiators or recipients of such mountings. Three mounts of adult males by two-year-olds were attempted and all received vigorous rebuffs. Adult males reacted differently when mounted by yearlings; they remained passive in 11 of the 12 instances. The reaction of the mounted animal appears to vary with the relative size and activity of the mounter: a large, vigorously thrusting animal generally elicits strenuous escape reactions from the one being mounted, while a small animal or a relatively passive large one is generally tolerated.

As the breeding season progresses, adult males and females become less closely associated. I evaluated male/female spatial proximity among hoary marmots during the first 4 weeks following emergence from hibernation by estimating the distance between male and female at 10-minute intervals. The average distance during weeks 1–2 (22.3 meters, SD = 6.4) was significantly less than during weeks 3–4 (46.1 meters, SD = 15.2).[17] Comparable results have been reported for hoary marmots in Alaska (Holmes, 1984b).

Marmota broweri, inhabiting the Brooks Range along Alaska's North Slope, apparently breeds within its den, prior to spring emergence (Rausch and Rausch, 1971), as do the Asiatic *M. camtschatica* (Kapitonov, 1960) and *M. sibirica* (Bibikov, 1967). *M. marmota* breeds after emergence (Muller-Using, 1957), as do the other North American species. Mating behavior among yellow-bellied marmots is characterized by male approaches to the female, female avoidance very early and late in the mating season (Armitage, 1974), and occasional female sexual solicitation primarily during the second week after emergence (Armitage, 1965; Fig. 6.36). The different flavor of yellow-bellied as opposed to Olympic or hoary marmot social life is apparent from these data, in that such categories as "female approaches male" or "male approaches female" would have little utility in describing sexual behavior in the latter two species, since avoidance is less prominent and approaches of one sort or another occur so frequently that they would not usefully distinguish sexual behavior from daily social life. By contrast, approaches by either male or female yellow-bellied marmots toward other conspecifics are unusual enough to be considered a special part of mating behavior.

Mating behavior among woodchucks has not been well described.

[17] t test, P < .01

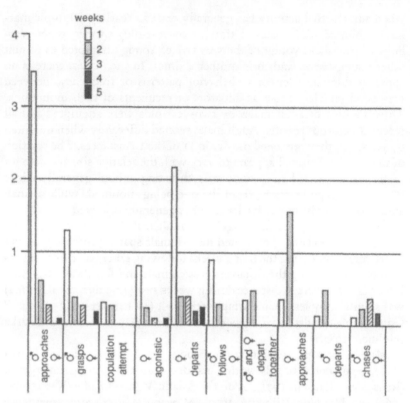

Fig. 6.36. Reproductive behavior of yellow-bellied marmots (*M. flaviventris*) during the first 5 weeks post-emergence. (Data from Armitage, 1965)

The annual cycles of testosterone production and of testis recrudescence and regression are known to be correlated, however, and at least in New York State, both are at maximum during the short late-winter breeding season (Baldwin et al., 1985). My observations suggest that adult males cohabit with adult females for 1–5 days shortly after the latter emerge from hibernation, after which the males typically move on, searching for other females, while the impregnated females remain resident in their burrows. Thus, I noted the development of early-season domestic arrangements of 14 different and apparently successful male woodchucks: nine of them ultimately established a residence within 0.5 km of "their" females, although virtually no interactions between these "mated pairs" occurred during the rest of the year. The remaining five adult males cohabited with a second female for approximately 2 days each. I never observed a female to cohabit with more than one male during the breeding season, either simultaneously or sequentially.

It is quite possible that post-estrus aggressiveness by female woodchucks drives the male away: I observed six fights, initiated by females and directed toward males; five occurred either the day prior to the male's departure or on the day of his departure. Because of the low degree of sexual dimorphism in body size among woodchucks, female aggressiveness may well be a significant consideration for the males. Just as the lack of simultaneous polygyny among woodchucks could be due proximately to female/female competition and intolerance, the occurrence of sequential polygyny in *M. monax* may be stimulated proximately by female intolerance of the male, if female aggressiveness increases as estrus ends and pregnancy begins.

In the absence of good field techniques for establishing paternity, it is difficult to assess the variance in male reproductive success among any species of marmots. However, if female woodchucks mate typically with one male per estrous cycle, with the association terminated by female aggressiveness resulting from an end of estrus and the onset of pregnancy, and if some males mate with more than one female, then it seems likely that some males do not breed at all. In addition to the 14 apparently successful adult male woodchucks described above, I also observed five adult males who never seemed to associate with a female. Female woodchucks, like their montane counterparts, appear to have rather brief periods of estrous receptivity. It would therefore seem that in this case, male variance in reproductive success is reduced over what it would be if estrus were more prolonged or if monopolization of mating opportunities were facilitated by a clumped distribution of females. This is because under these conditions it would presumably be easier for a small number of males to monopolize the available females. To some extent this may well occur, as suggested by the relatively low level of sexual dimorphism in this species. However, even though individual adult female woodchucks seem to have brief estrus, the estrus of different females is not apparently synchronized, and as a result there is a wide seasonal range during which some females within any local population are potentially reproductive.

The woodchuck's mating season may be three weeks or more at southerly latitudes (Grizzell, 1955; Snyder and Christian, 1960), briefer at higher latitudes. The wide range of woodchuck mating seasons is due both to individual variability in the post-hibernation emergence times of different females and to the facts that 20–50% of yearlings breed in this species (Snyder, Davis, and Christian, 1961) and yearlings emerge from hibernation several weeks later than adult females. Of the five sequentially polygynous adult male woodchucks described above, the second mates of three were yearling females. Male woodchucks are relatively nomadic for the first 6–8 weeks after emergence, traveling through a wide home range, within which they cohabit with receptive females. The longer above-ground period of activity enjoyed by woodchucks has probably relaxed selection for brevity of mating season, which in turn seems to have led to occasional sequential polygyny.

By contrast with the woodchuck's situation, the montane species are much more sedentary—some occasional "gallivanting" notwithstanding (see Chapter 8). Among the colonial marmots, one or more females are concentrated into a single colony with intervening habitat typically not congenial for wandering individuals. Moreover, female estrus is more closely synchronized in the montane species, concentrated in a single week rather than spread through several weeks as in *M. monax*. This is presumably because of the short above-ground season and the consequent necessity to mate early if young are to gain enough weight to hibernate successfully.

Although the extended mating season of woodchucks provides the potential for a high variance in male reproductive success, this effect may be countered by the greater physical spacing (resulting from individual intolerance) in this species. By contrast, the more limited breeding season in the montane forms would seem to encourage more evenly distributed male reproductive success, while at the same time this effect is countered by the clumped distribution of adult females. Since sexual dimorphism appears to be greater among the montane species, it seems likely that on balance the variance in male reproductive success is also greater among these species. However, since numerous other factors might also be responsible for sexual dimorphism, this conclusion must be advanced cautiously.

Summary and Conclusions

1. "Greeting" is a frequent behavior among marmots, involving face-to-face interactions that presumably serve individual recognition.
2. Greeting is most frequent among Olympic and hoary marmots, less so among yellow-bellied marmots, and rare among woodchucks.
3. Greeting varies with age and sex class, as well as with season, and it characterizes changes in daily social patterns.
4. It may be misleading to consider all greeting "amicable," although

it clearly varies directly with the level of social interaction and serves as a useful means of contrasting the behavior of different individuals and different species.

5. Play-fights are also prominent parts of the marmot behavioral repertoire, involving upright postures and apparently ranging from play (when performed by younger animals) to fighting (when done by older ones).

6. Play-fighting varies seasonally, across age/sex categories and across species, being most common among younger animals and among the colonial, montane species.

7. Vocal communication is especially well developed among marmots, the most prominent of which involve variations on a high-pitched tone associated with predators or agitation.

8. Olfactory communication is presumably exchanged at greetings, and cheek and anal glands are also employed.

9. Visual communication is achieved by use of the conspicuous, upraised tail.

10. The Vancouver marmot is distinguished from other species studied thus far by several aspects of its communication: notably, use of a distinctive call and the use of cheek glands for territorial marking.

11. The mating behavior of marmots is relatively simple, characterized by various patterns of male approach to the female and responses by the female, including avoidance and receptivity. Electrophoretic analysis has not yet been developed for marmots, however, and would be a welcome source of information, including confirmation of paternity.

12. Mating occurs shortly after emergence in the spring, and although effective mating behavior could well take place underground, enough is observed above ground to permit description. More data are needed, however, on the timing of breeding as well as the actual behavioral patterns of different species.

13. Sequential polygyny seems to be the rule for woodchucks, although it is uncertain how many males succeed in establishing multiple mateships; female intolerance of the male may be a proximate cause of the solitary postbreeding living arrangements of woodchucks.

14. Longer breeding seasons (as among woodchucks) as well as earlier sexual maturation of females provide a greater opportunity for variance in male reproductive success; by contrast, the more restricted breeding season of the montane species would seem to encourage lesser male variance in reproductive success, an effect which, however, may be countered by their polygynous living arrangements.

CHAPTER 7

Behavioral Ontogeny and Individual Differences

S ocial behavior arises from interactions among different individuals. But it does not simply result from each encounter taken alone, like the spark produced when two objects clash together. Rather, the patterns of animal (and human) sociality also have a developmental component: they are influenced by the previous experience of each animal involved. And perhaps the most crucial of such experiences are those that take place during the development of each individual. In this chapter we therefore examine the ontogeny of marmots, paying particular attention to behavior. Furthermore, since each individual is genetically distinct, and personal experiences are also idiosyncratic, it seems likely that individual differences in behavior are inextricably tied to the ontogenetic trajectory of each individual. We therefore conclude this chapter with a brief examination of behavioral individuality among marmots.

Behavioral Ontogeny

The general marmot pattern is for gestation to last about 4 weeks, and weaning another 28–30 days, up to 44 days among woodchucks. The extended weaning period of *M. momax* may be due to a relaxation of seasonal constraints, since young montane marmots should be under especially intense pressure to accumulate sufficient size and fat during the restricted above-ground period and nursing is generally less effective than foraging as a way of gaining nourishment. On the other hand, nursing young are probably less subject to predation. Moreover, by delaying weaning they might delay the onset of female intolerance and, hence, their own dispersal.

Fig. 7.1. Adult female yellow-bellied marmot (*M. flaviventris*) nursing her young. Among all species of marmots, nursing only rarely occurs above ground. (Photo by K. B. Armitage)

Young marmots of all species typically are not seen above ground until they are being weaned, and, in fact, weaning and the resulting hunger may well be the proximate stimuli that induce the young marmot to emerge from its natal burrow. Upon emergence, the young are fully furred and ambulatory (Fig. 7.1).

For about ten days immediately preceding the emergence of their young, Olympic marmot mothers spend considerable time near their littering burrow, making only brief feeding sorties and typically returning within 20 minutes. They may spend little time within these burrows, however, often lying down on the dirt porch just outside. When the young first appear, they make only very brief excursions. After 4 days they are generally out for less than 10 minutes at a time and always within a meter or two of their burrow (Fig. 7.2). From then on, outside time as well as distance traveled from the burrow increases until by about 2 weeks after emergence they remain out nearly as long as any adult (Fig. 7.3). However, they generally stay within 10 meters of their home burrow for at least another week. The radius of the young marmot's world expands regularly, with the natal burrow as center, until in about 3 weeks it encompasses the entire colony (Fig. 7.4).[1]

[1] Mann-Whitney U tests, $P < .05$ for days 4 vs. 8, and 8 vs. 23

Fig. 7.2. Two young Olympic marmots (*M. olympus*) within a few days of their first emergence, on the dirt porch by their natal burrow. (Photo by D. P. Barash)

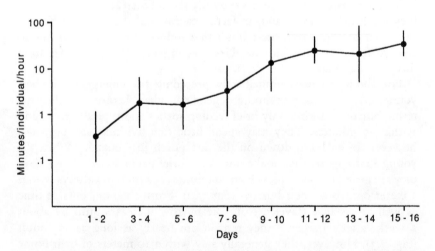

Fig. 7.3. Time spent above ground by *M. olympus* young of the year as a function of days post-emergence. Data are presented as minutes above ground per animal per observation-hour, based on 87 observation-hours of five different young.

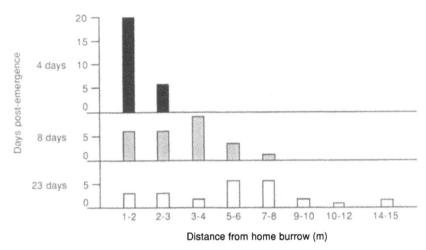

Fig. 7.4. Distance of *M. olympus* young from their natal burrow as a function of time post-emergence. Each datum is the mean distance for three young; distances recorded every 15 minutes for 7 hours each day.

At about 1 week after emergence, young Olympic marmots begin visiting other burrows, although they remain more wary than yearlings or adults and are more likely to run to their burrows in response to novel occurrences. They begin foraging almost immediately after emergence; reproductive females' mammaries are substantially reduced within 2 weeks of their offspring's appearance, and weaning is almost certainly complete within 3 weeks after that. No above-ground suckling was ever observed for Olympic or hoary marmots.

The behavior of the mother also changes following the appearance of her young: she spends progressively less time in the immediate area of her burrow and more time away from it, while time spent within the burrow remains essentially unchanged (Fig. 7.5). During the first 2 weeks, the mother feeds at increasing distances from her littering burrow, returning to the young at increasing intervals until they are left alone virtually the entire day (Fig. 7.6). When not feeding, young marmots spend most of their time playing, often vigorous rough-and-tumble play with much prancing and pauses for mutual grooming and predator avoidance (Fig. 7.7).

After a few days, young Olympic marmots are greeting and being greeted by all other colony members, although their emergence does not elicit any obvious new behavior by the other residents. These animals

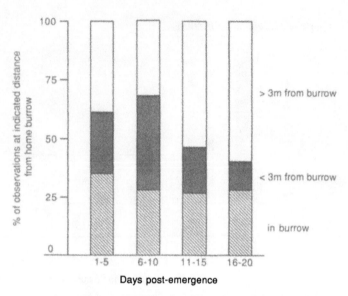

Fig. 7.5. Distance of reproductive female Olympic marmot *M. olympus* from her burrow as a function of time post–emergence. Distances were recorded every 5 minutes during 15 hours of observation during each 5-day period.

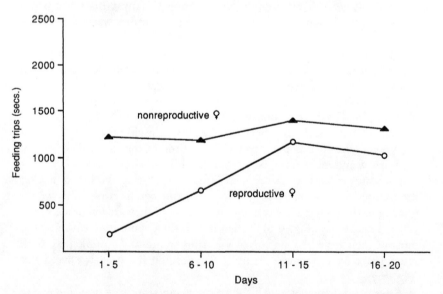

Fig. 7.6. Duration of foraging trips by reproductive and nonreproductive adult female Olympic marmots (*M. olympus*) as a function of days following emergence of young at their colony. Based on a minimum of eight different foraging trips by each of three different females in each case.

Fig. 7.7. Three young Menzbier's marmots (*M. menzbieri*), shortly after weaning. (Photo by A. Nikolskii)

had freely visited the natal burrow before the young were weaned, and presumably socialization is already well advanced at emergence. Other adults and yearlings seem to make no particular effort to encounter or to avoid the newcomers, although for about the first 2 weeks, the young conspicuously avoid the adult male, returning to their burrow whenever he approaches; however, the adult male typically displays no obvious aggression toward them.

M. olympus colony members are extremely tolerant of the young, who frequently crawl over the larger animals, often chewing their fur. Older animals also submit freely to extensive facial mouthing (resembling a long, sloppy greeting), which is not long tolerated if initiated by an older animal. For example, 19 (83%) of 23 yearling-initiated lengthy greetings of adults in June elicited a reprimand from the adults, whereas only four of 18 (22%) lengthy juvenile-initiated greetings of adults were followed by an aggressive adult response from the adult.[2] The frequency of early-season extensive facial attention by yearlings declines with the season, and with it the frequency of aggressive adult responses.

The decrease in mouthing of adults by yearlings may reflect matura-tion and/or learning by the yearlings—learning that such behavior, tol-

[2]binomial test, $P < .01$

erated in them as young of the year the previous summer, now elicits reprimands from the adults. The adults' intolerant reaction to yearlings' mouthing, compared with their indulgence of the young, may be precipitated by the greater size of the yearlings or by the fact that such mouthing occurs early in the season when adult aggression is already high. Certainly, by the emergence of young of the year, seasonal aggressiveness has waned and greater adult tolerance (presumably related proximally to hormone changes) is added to whatever inhibitions the young themselves may provide.

Behavioral ontogeny among hoary marmots appears quite similar to that of Olympic marmots, with one peculiar exception: I have seen four different mother hoary marmots transport their young, carrying them one at a time in the mouth, and depositing them at distances of up to 30 meters from the natal burrow. This occurred 13 times altogether (one female moved her young five different times). In all cases this occurred during weeks 1–3 following emergence of the young. Interindividual distances increase progressively among newly emerged young hoary marmots, from a mean of 3.7 meters (SD = 3.2) during week 1 after emergence, to 17.2 meters (SD = 5.3) during week 2, to 27.6 meters (SD = 9.9) during week 3.[3]

Among hoary marmots, greetings and grooming may occur from the first day of emergence; first play-fights were seen on day 3, and upright play-fights were seen on day 5. Squeaks and high-pitched squeals occur from day 1. The first alarm call was given by a young animal on day 5, although young hoary marmots respond to alarm calls from day 1 (see also Holmes, 1979). Hoary marmot mothers appear to interact somewhat less with their offspring than do Olympic marmot mothers: the former averaged 0.35 greetings initiated with their young per female per observation-hour (SD = 0.12), whereas the latter averaged 0.88 (SD = .25).[4] The biological significance of this difference is unclear, but it may relate to the different physical topographies normally experienced by the two species. Since Olympic marmot burrows tend to occur in grassy meadows whereas hoary marmots preferentially occupy rockpiles and talus slopes, the latter may simply experience more physical interruptions in combining their own foraging with attention to their offspring. Hoary marmot mothers often lie down near their young at the natal burrow, although even then they appear to interact less frequently than do Olympic marmots.

[3] Pearson's product-moment correlation, $P < .01$
[4] Mann-Whitney U test, $P < .05$

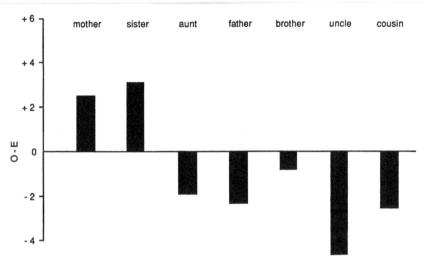

Fig. 7.8. Entering of natal burrows by hoary marmots (*M. caligata*) as a function of the genetic relationship between the burrow enterer and the reproductive female occupying the burrow. Data are presented as Observed − Expected, calculated as suggested by Altmann and Altmann (1977); based on a minimum of 11 different dyads and a minimum of 125 observation-hours for each category.

As with Olympic marmots, it seems likely that socialization of young hoary marmots begins before weaning, since colony residents enter the natal burrow prior to seasonal emergence of the young of the year. Because marmot social behavior is generally a nonrandom function of kinship, it is particularly interesting that during the presumed preweaning socialization period, close genetic relatives of the young marmots are more likely than distant relatives to enter natal burrows (Fig. 7.8).[5] If social preference and tolerance results in part from early experience, the pre-emergence experiences of these animals may well be a proximate cause of their subsequent prosocial behavior toward relatives, especially members of the same matriline. Such tendencies would also be perpetuated in succeeding generations insofar as adult females, socially exposed to their relatives from infancy—and hence predisposed toward them— differentially facilitated access by these same individuals to their own offspring.

Young yellow-bellied marmots are less socialized than young hoary or Olympic marmots. Although yellow-bellied marmots engage in substantially less play than their high-elevation counterparts, play nonethe-

[5] chi-square test, P < .01

less constitutes more than 40% of social interactions among young of this species (Nowicki and Armitage, 1979). Young male yellow-bellied marmots initiate greetings with their mother significantly more than do young females (Nowicki and Armitage, 1979). By contrast, no sexual difference of this sort can be detected among young hoary marmots. Similarly, young male yellow-bellied marmots devote on average nearly twice as much time to chasing and sexual play as do females (Nowicki and Armitage, 1979), whereas no difference occurs among hoary marmots. This comparison is consistent with behavioral development in these two species, since young yellow-bellied marmots are only one year away from dispersing and two years from sexual maturity, whereas young hoary and Olympic marmots are two years from dispersal and three years from sexual maturity. Delayed maturation correlates with delayed attainment of behavioral dimorphism among the young of the year.

For their part, young woodchucks, as expected, apparently assume adult behavior patterns even more rapidly than do yellow-bellied marmots. Thus, I have observed active rough-and-tumble play among free-living woodchucks only during the first and second days after their emergence. I have never observed individuals other than littermates or the mother enter a woodchuck natal burrow. By day 6, young woodchucks forage independently of each other and of their mother. Their only consistent social interactions take place briefly at the entrance of the burrow just after daily emergence and before the animals begin foraging. However, young woodchucks occasionally solicit interactions with older animals, generally by a tentative approach, seeking to initiate a greeting, or (more subtly) failing to avoid an adult. In such cases, the young animal is most commonly ignored, sometimes threatened, and rarely lunged at. Dispersal typically occurs between the first and the third week; by the end of their first week above ground, young woodchucks are showing much of the independence and social intolerance that characterizes adults, and their above-ground experience with other adults—notably their mother—may contribute substantially.

Prior to their emergence, the treatment of young woodchucks by their mother is characterized by a progressive decrease in maternal solicitude and a corresponding increase in maternal intolerance (Fig. 7.9). Preweaning behavior of the montane marmot species has not yet been studied; it therefore remains an open question to what extent—if at all—patterns of social tolerance and intolerance among adults of the different species are prefigured by maternal interactions with weanlings. Free-living female woodchucks generally remain underground for a few days before and

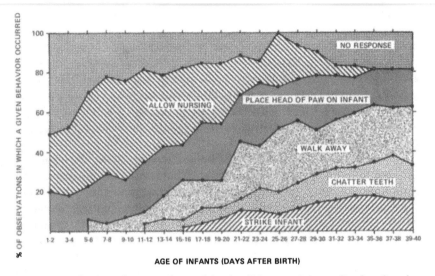

Fig. 7.9. Behavior of maternal woodchucks (*M. monax*) immediately after the onset of squeaking by the young. Based on data for three litters, a total of 387 incidents and a minimum of 14 per 2-day interval; 127 observation-hours.

after giving birth, and captive animals are "very belligerent" for several days after parturition (Grizzell, 1955).

During their first year, young woodchucks tend to occupy smaller home ranges than do adults. I sampled 20 home-range measurements for each of eight different young of the year in central New York State during July, and found a mean of 1,115 square meters (SD = 223), whereas 20 comparable measurements on eight different adults averaged 2,314 square meters (SD = 488).[6] By contrast, home ranges determined similarly for six different yearlings averaged 2,010 square meters (SD = 522), which is not significantly different from the home range of adults.

Individual Differences

In the past, biologists often did not distinguish among the ages or even the sexes of their subjects. Old accounts of marmot behavior (by Ernest Thompson Seton, John Muir, John Burroughs, and C. Hart Merriam) frequently omit whether the animals in question were male or female, young or old. As research on animal behavior and ecology has taken on

[6]t test, P < .01

an increasingly evolutionary flavor, such considerations have become paramount, and indeed, differences between male and female and between various age classes are a major focus of research and analysis, complete with well-defined theoretical expectations. Future researchers, however, may well find current studies as incompletely specified and, therefore, curiously unsatisfying, as we sometimes find the work of nineteenth-century naturalists. Just as the latter often ignored age and sex differences, modern-day sociobiologists often ignore the *individual differences* among their subjects.

To some extent, this tendency is understandable. After all, insofar as one goal of research is to understand the differences among the behavioral strategies of adults, yearlings, and young, between satellite and colony males, between reproductive and nonreproductive females, between males and females, or, for that matter, among various species, grouping individuals is desirable and, indeed, necessary if we are to draw generalizations about the larger entity of which they are a part. We typically take some convenient statistical abstraction such as the mean or median to represent these larger entities, despite the fact that in the process of compressing and condensing our data, we also ignore the possible role—even the existence—of differences among individuals. Sociobiology has not yet begun to articulate a coherent theory of individual differences. Some day, perhaps, it will.

Notable efforts at assessing individual behavioral profiles have been conducted in studies of yellow-bellied marmots. The technique employed has been "mirror-image stimulation," which essentially involves exposing captive animals to a mirror, and recording their subsequent behavior (Svendsen, 1974; Svendsen and Armitage, 1973; Armitage, 1982b).

Analysis of the results obtained from mirror-image stimulation revealed that 85% of the observed variance in behavioral responses could be explained by three factors, and that individuals could reliably be classified into one of three categories defined by the dominance of these factors. Thus, animals characterized by "approach" spend most of their time oriented toward the mirror; "avoiders" do not make contact with their image and tend to sit far away from it; finally, "sociable" individuals nose and paw their image, tail-wag, and often engage in other activities, such as eating or exploration. Unfortunately, this technique suffers from the disadvantage that each individual necessarily receives a different stimulus: an adult male sees himself in the mirror, a young of the year sees itself, etc. Aggressive animals see aggressive animals, submissive animals see submissive animals, and so forth.

Nonetheless, mirror-image stimulation has revealed suggestive and potentially important patterns relating individual behavioral profiles to the normative behavior of free-living *M. flaviventris*. Behavioral profiles obtained in this manner correlate with the social situation of the free-living animal.

As Svendsen (1974) described it, "Aggressive females tended to occupy small harems at large sites, or to live as solitary individuals regardless of the size of the site. Social females occurred primarily at large sites, in harems. Avoider females lived in peripheral burrows at colonies, or in satellite sites. All behavioral types reproduced successfully, but reproductive fitness was correlated with social structure. Aggressive females were most fit in small harems or at satellite sites, whereas social females were most fit in large social groups. Avoiders were the least fit in all social situations." The behavior patterns identified via mirror-image stimulation were also stable over time and repeatable (Svendsen and Armitage, 1973).

I assessed unidimensional profiles of the aggressiveness of 5 adult female hoary marmots by characterizing them as to ratio of chases initiated to chases received before live-trapping them and introducing them into new colonies. Following their transplantation, I reassessed each female and found the following correspondences (chase ratio before transplantation / chase ratio after transplantation): 0.48/0.35, 0.79/0.94, 1.1/0.9, 1.3/1.1, and 2.2/0.8. Although there was a consistent (and not surprising) trend for these females to show lower chase ratios after being transplanted, the most interesting finding is that in all but one case, profiles of aggressiveness also remained consistent: aggressive individuals tended to remain relatively aggressive and submissive individuals tended to remain relatively submissive. Interindividual differences were significantly greater than "intraindividual" differences.[7] A similar pattern was revealed when nontransplanted adult males were compared in successive years, and also when nontransplanted adult females were contrasted in years of comparable reproductive status—i.e., nonreproductive females compared to their next nonreproductive year, and contrasted similarly with other nonparous females. Reproductive females, by contrast, showed a relatively low "treatment effect" between individuals, presumably because a more relevant treatment effect (pregnancy) tended to equalize individual differences.

In any event, it seems clear that marmots have individual behavioral profiles, and that these profiles are relatively constant from year to year. Regardless of the optimum technique(s) for assessing individual differ-

[7]Kolmogorov-Smirnov one-way analysis of variance, $P < .05$

ences in marmot behavior, it seems clear that such differences exist, and that when individuals of the same social and biological role are considered, the differences between individuals are greater than the differences "within" individuals, and this persistence of marmot personality may well have important correlates and consequences.

For example, satellite males may be more likely to establish themselves when the colony male is relatively unaggressive (Chapter 8). Aggressive adult females may be more likely to inhibit the reproduction of other females (Chapter 9), and aggressiveness by adult males and/or females may influence the pattern of dispersal, and hence the basic social structure of their colonies (Chapter 13). The behavioral tendencies of individuals resident in a particular colony may strongly influence observed patterns of social interaction, such that colonies occupied by aggressive individuals may be characterized by relatively high frequencies of chasing, fighting, and mutual avoidance. Such considerations may bedevil efforts to recognize general patterns of social organization. For example, the considerable variability reported for yellow-bellied marmot colonies (Armitage, 1977) seems to be influenced by such factors as the following: (1) whether there are many immigrants present (the presence of immigrants generates a higher level of agonistic behavior than when the residents interacted together the previous year (this is especially true of newly arrived males); (2) the age/sex structure of the population (e.g., if yearling males are present, agonistic behavior is more likely); (3) population size (increased colony population leads to more agonistic and less affiliative behavior); (4) the partitioning of space (for example, if a particularly dominant female's home range overlaps that of many others, agonistic behavior is likely, whereas if the home ranges of "sociable" females overlap, affiliative interchanges are more frequent, although the frequency of agonistic behavior may also be high, simply because of enhanced proximity or, more appropriately perhaps, reduced avoidance); and finally, (5) the individual characteristics of residents may have a profound effect, such that a single highly aggressive animal, for example, could be disproportionately influential. Although nonparametric statistical tests can adjust for extreme behavioral data, there is no appropriate way to adjust for the actual effects of behavioral extremism, since these individuals interact with others, producing additional effects throughout the colony.

The numerical density of individuals is typically taken as the relevant index of population density, whereas "behavioral density" may be a far more appropriate measure (Armitage, 1975), although one that is difficult to determine. Behavioral density could be high even at low numerical density if the individuals in question are inclined to interact frequently because of their individual behavioral tendencies and inclinations. And vice versa, a population with high numerical density may actually experience low behavioral density if the component individuals are likely to avoid one another, either because of a high frequency of "avoider" personalities or because one or more aggressive individuals causes the others to maintain greater interindividual distances.

As we shall see (Chapter 13), the behavioral profiles of resident adults may well influence tendencies for dispersal (Armitage, 1975, 1984). Although the dispersal of yearling female *M. flaviventris* does not seem attributable to aggression from the adult male, adult females have been seen to chase female yearlings from their colonies. Accordingly, the existence of a "sociable" adult female appears to increase the retention of female yearlings, and nondispersing female yearlings show a strong tendency to overlap the ranges of "sociable" adult females (Armitage, 1975, 1977, 1984).

Tolerant adult females presumably suffer the disadvantage of greater female/female competition but might also enjoy kin-selected benefits via the potentially greater fitness of their nondispersing offspring. Of course, we also expect that individual behavioral tendencies like aggressiveness are modifiable, depending on such circumstances as forage quality, the nature and number of nearby relatives, reproductive condition, etc. And if there are fitness consequences to different patterns of behavioral individuality, then we can predict that selection will operate on tendencies toward behavioral individuality so as to maximize inclusive fitness, no less than it operates on other phenotypes. Thus far, however, no satisfying relationship has been demonstrated between individual differences among yellow-bellied marmots (as determined by mirror-image stimulation) and reproductive success; no significant correlations exist among "approach," "avoidance," and "sociability," and such measures as number of weaned young, number of yearlings, or number of yearlings successfully recruited (i.e., nondispersers) into the natal colony (Armitage, 1986a). This might suggest that individuality occurs as a number of different types with variance around each type, rather than as a single, continuously varying phenomenon.

If we assume, for the sake of argument, that individuals can be characterized accurately as "aggressive" or "sociable," to some degree independently of their living circumstances, then a population composed primarily of sociables would be susceptible to invasion by aggressors. As the aggressors drove out their less dominant colleagues, colonies composed of aggressive individuals would then be immune to replacement by sociables. However, since aggressors would be likely to establish small colonies, they would receive fewer of the advantages of sociality *per se*, such as predator avoidance, biological conditioning of their environment, etc.; hence, such colonies might be short-lived in comparison to those composed of sociable individuals. The smaller size of such colonies would in itself make them more likely to become extinct. As a result, even though aggressor personalities might seem to be a pure Evolutionarily Stable Strategy (because they would be resistant to invasion), substantial heterogeneity could still be maintained.

Individual differences in behavioral profiles can nonetheless be expected to have profound demographic consequences, if tendencies to disperse vary with temperament, such that aggressive individuals tend to induce the dispersal of other, relatively intolerant adolescents and also prevent the immigration of others. Moreover, sociable individuals seem less likely to modify their own spatial patterns after immigrants succeed in establishing themselves and/or after others reproduce, whereas avoiders might respond by becoming peripheral or perhaps even dispersing altogether. In this regard, the differential impact of individual variability in the various marmot species is difficult to predict: for example, the more socially integrated Olympic and hoary marmots would presumably experience the effect of diverging behavioral profiles more intensely than would the more spatially dispersed yellow-bellied marmots, while woodchucks should be the most indifferent and insensitive to such variations. On the other hand, the fact that a species experiences a high level of social integration may also suggest precisely the opposite, namely, that such species are relatively unaffected by individual social vagaries.

It should be pointed out, however, that speculation such as this must be considered tentative until better data are available. Moreover, there seems to be little if any correlation between the behavioral profiles of adult females—as determined by mirror-image stimulation—and those of their offspring (Svendsen, 1974), although a full range of behavioral responses is present among young of the year. The patterns of behavioral interactions among young yellow-bellied marmots seem to be strongly influenced by individual behavioral profiles, which in turn appear to be more influential than patterns of genetic relatedness *per se* (Armitage, 1982b). Similarly, the greatest part of the variance in female reproductive success is explained by variance among individual females, rather than by the variance among colonies or among matrilines (Armitage, 1986b).

When individual differences in behavior are evaluated, several points should be kept in mind. For one, individual differences should be distinguished from differences based on different biological and social roles.

Satellite male hoary marmots are consistently subordinate to the colony male: this difference between the two males is, in a sense, an important aspect of their behavioral individuality, and differences between colony and satellite males are almost certainly responsible for their very different social situations and, therefore, the observed distinguishing patterns. However, when and if satellite males eventually replace colony males, they become colony males themselves, whereupon their behavior typically becomes that of colony males generally. Reproductive females spend more time near their burrows, interacting with their young, and they are generally more dominant than nonreproductive females. Such differences reverse themselves the following year when their reproductive roles are reversed. Accordingly, it seems most useful to restrict the concept of behavioral individuality to distinctions between individuals that are socially and biologically as similar as possible in all other respects, notably age, sex, social status, physical health, residence situation, and reproductive state.

In addition, a distinction must be made between intraindividual variability (the variability demonstrated by a single individual under different circumstances and/or different stages of ontogeny) and interindividual variability (variability among different individuals). The former refers to the plasticity of each individual's behavior, which should legitimately include both variability from one situation to another within a brief time span as well as variability from year to year. Increases in intraindividual variability also add to the total observed interindividuality. For example, most marmots tend to become increasingly dominant with age; this consistent pattern increases—or at least maintains—the variability obtained within any one colony since, as a result of the overlapping generations characteristic of marmot life, individual variability among adults is to be expected. To be consistent with the point made above, therefore, any measures of observed behavioral individuality should compare like-aged individuals in all cases. Because of the frequent paucity of data for free-living populations, this is often difficult to achieve, and typically "adults" are compared, just as "two-year-olds," "yearlings," or "young of the year" are.

Finally a distinction must be made between consistent differences between individuals—each of which might remain quite stable in its behavior—and fluctuations in the behavior of any one animal, which produces ever-changing differences between individuals.

The following generalization can be suggested for age-related behavioral variability. It appears to hold for marmots, and may have greater generality as well. It is really an empirical generalization, rather than a theory or even a hypothesis, since it is based simply on observed patterns, not on any underlying conceptual structure. It is, however, refutable. The generalization is simply this: with increasing age, marmots show a decrease in intraindividual variability and, at the same time, an increase in interindividual variability. As individuals mature, their behavior becomes more consistent ("more like themselves") while simultaneously becoming more distinct from that of others ("less like others"). The behavior of young individuals is generally more variable and less predictable than that of older ones, which become more differentiated as they mature and, of course, as they assume more distinct social roles. This pattern of differentiation of each individual resembles a multidimensional cone, whose base represents young of the year. If different individuals are considered together, their behavioral cones tend to overlap substantially at their bases, then point in somewhat different directions as the individuals' behavior differentiates with age (Fig. 7.10).

Consistent with this model, mirror-image stimulation among yellow-bellied marmots revealed that young animals are generally not suffi-

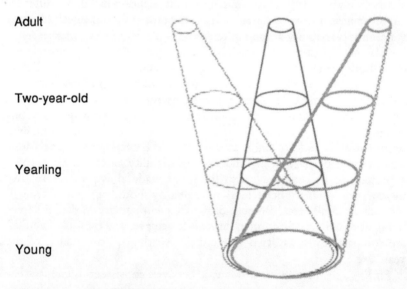

Adult

Two-year-old

Yearling

Young

Fig. 7.10. Conceptual scheme of the ontogeny of behavioral individuality among marmots. Each hypothesized cone represents the behavioral envelope for a different individual, progressively narrowing and overlapping less with others as age increases.

ciently differentiated from each other to be assigned to the behavioral profiles "avoider," "aggressive," or "social," appropriate for older animals (Nowicki and Armitage, 1979).

And in another study, young animals classified as "sociable" did not play, whereas "avoiders" did so significantly more than expected; in addition, "sociable" young greeted less than expected from the number in the sample, whereas "avoiders" did more than expected (Armitage, 1982b). In short, categories of behavioral individuality for young yellow-bellied marmots appear to be less distinct than for older animals.

Behavioral individuality among European Alpine marmots (Barash, 1976a) suggests a similar pattern. Individual differences among adults exceeded those of yearlings; young of the year, unfortunately, were not studied as individuals. Between-colony individual differentiation exceeded differentiation within colonies. It is also interesting to note that comparatively asocial maintenance behaviors—foraging time, distance from the home burrow, frequency of self-grooming, and frequency of alarm calling—revealed no age-related trends in behavioral individuality. This could be due to lesser canalization of the genetic aspects of social behavior; alternatively, such asocial behaviors as time spent foraging and distance traveled, as well as frequencies of self-grooming and alarm-calling, may largely be direct functions of local environmental conditions. If so, then similarities in such cases may simply reflect ecological similarity of different sites. By contrast to maintenance behavior, social behavior may also be more susceptible to modification by individual experience and the idiosyncratic constraints imposed by others within a colony.

I determined individual behavioral profiles for *M. caligata*, using field observations alone, by recording the number of greetings per hour, play-fights per hour, and chases initiated per hour based on 10 hours' observation of each of 7 different four- and five-year-old colony males, 10 different four-year-old nonreproductive colony females, 10 different yearling males, and 10 different female young of the year. Data were obtained during the first 2 weeks in July, except for the young of the year, which were studied during the first week of August, in all cases between 0700 and 1100 hrs.

When individuals of the same age and sex class are compared, and the observed standard deviation for each individual is expressed as a single data point, standard deviations tend to vary directly with the means of the behavior in question.[8] An appropriate measure of variation therefore

[8] Kendall correlation coefficient, $P < .05$

relates the degree of variation in an observed trait to the magnitude of the trait itself—i.e., by determining the ratio of standard deviation to arithmetic mean. This yields what might be called a "relative standard deviation," a measure of the observed variation adjusted for the magnitude of the behavior in question. Taking this measure for each individual as a single data point, dispersion for the seven adult males was less than that for the ten yearlings, which in turn was less than that for the young (Fig. 7.11).[9] The ten nonreproductive females were not statistically distinguishable in this respect from the adult males.

Thus, with regard to the recorded behavior, older animals behave more consistently than do younger ones. By contrast, the higher standard deviations of younger animals indicate that their behavior, from hour to hour, is more variable. Hence, the behavioral cones, depicted in Fig. 7.10, are wider at the base (younger animals) than at the apex (older animals).[10] It should be noted, however, that the standard deviations of some behavioral measures—an index of statistical dispersion—must be distinguished from mean or median, an index of the amount of the behavior in question. Thus, adult males, for example, can have a low standard deviation to mean ratio for greeting while also having a high average greeting rate.

The above treatment suggests that older animals are more consistent (show less intraindividual variability) than do younger ones. Analysis of variance, using the data themselves rather than a transformation of those data, revealed that interindividual differences are greater than intraindividual differences for adult males and adult females[11] but not for yearlings or for young of the year. To summarize and simplify the above findings: the first part suggests that the behavioral profiles of individuals are somewhat cone-shaped—broader at the base when the individual is young, and narrowing as each individual matures. In other words, younger animals are less consistent than older ones. The second part suggests that the ontogenetic cones tend to point in different directions, diverging from each other as each individual matures and becomes increasingly differentiated; older animals are not only more consistent individually than younger ones, each individual as it grows older also appears to become more different from others.

Increasing behavioral individuality with increasing age is not surprising. It could be due to the accumulating effects of individual experiences, the result of simple maturation of differing genotypes, or more likely, a

[9] Mann–Whitney U tests, $P < .05$
[10] F-max tests, $P < .05$ for young vs. yearlings and for yearlings vs. adults
[11] $P < .05$

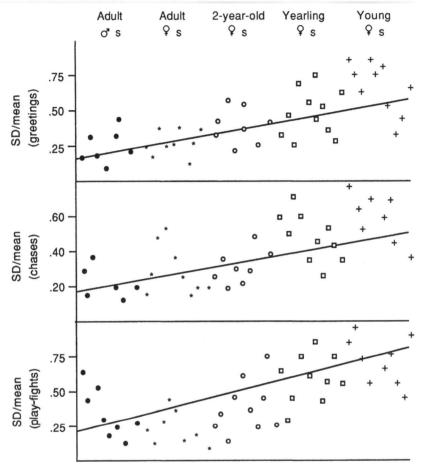

Fig. 7.11. Ratio of standard deviations to means for greetings, chases, and play-fights among hoary marmots (*M. caligata*) of different sex and age classes. Each datum represents a single individual, based on 3 years' data and a total of 217 observation-hours with a minimum of 8 hours' observation of each individual.

combination of these. The greater the age of each animal, the greater the opportunity for each distinct genotype to be expressed and for idiosyncratic personal experiences to widen the phenotypic gap between individuals. Even intra-uterine positioning can influence phenotypic variation among rodents (vom Saal, 1981), and the interactive patterns among littermates are not only a function of existing differences between individuals, but also a cause of such differences (e.g., Bekoff, 1977).

There are no data at present concerning behavioral individuality among woodchucks, and indeed, current theory does not yield predictions as to the relative salience and patterning of individual differences among solitary vs. social species. Since the latter tend to experience both delayed maturity and more defined social roles, they may also be expected to reveal greater behavioral individuality. Alternatively, since social species are more subject to the possibly homogenizing influences of other animals, it is possible that individual differences are more pronounced among solitary animals.

It also remains to be seen whether individual differences in behavior represent evolutionary adaptations *per se* and/or to what extent they are simply phenotypic "noise." Thus, the reality of meiosis combined with genic and chromosomal mutation can reasonably be expected to generate a degree of behavioral individuality no less than biochemical or anatomical individuality. Moreover, insofar as heterosis confers a fitness advantage upon heterozygotes, the resulting maintenance of diversity can, in theory, generate behavioral morphs which may be less fit than their alternatives (i.e., "avoiders" among yellow-bellied marmots).

An inexperienced observer of marmots will find it difficult to distinguish one adult from another, even using physical characteristics. But, in fact, if you have seen one marmot, you definitely have not seen them all. Adults are readily identifiable on the basis of pelage and other gross physical traits alone. Among the Olympic marmots, in which adults experience a pronounced midsummer molt to a dark black outer fur, these adults—especially the males—are unmistakably differentiated, even to a casual human observer. It is more difficult to distinguish among two-year-olds, more difficult yet among yearlings, and when it comes to young of the year, even the experienced and dedicated marmoteer may be hard pressed to tell one individual from the next.

The physical and behavioral differentiation among older animals may well provide greater opportunity for individual recognition by the marmots themselves. This in turn could select for heightened behavioral differentiation of adults over yearlings, and of yearlings over young of the year, since recognition of individuals presumably assumes greater importance as the animal to be recognized becomes older; it is difficult—although not impossible—to conceive of situations in which individuals would profit by concealing their identities. (One exception could be the satellite males, who might gain by concealing their sex, but show no signs of doing so.)

Finally, insofar as the reproductive opportunities and constraints of being an adult necessarily require the assumption of more distinct and more differentiated roles than is the case for younger animals, maturity itself might simply produce greater behavioral individuality among the adults. Among relatively long-lived, sedentary species such as marmots, in which individuals interact regularly, there seems little opportunity for the sort of sexual mimicry or deception suggested for some fish (Gross and Charnov, 1980) or birds (Rohwer, Fretwell, and Niles, 1980).

The fact of behavioral individuality among marmots seems undeniable. Its origin—whether genetic, developmental, or (most likely) some interaction of the two—is unknown. Its significance is also unclear: it may be adaptive or merely a random variation due to the unavoidable vagaries of meiosis plus idiosyncratic individual experiences. There is currently no evidence for adaptive variance in personality profiles among marmots—that is, for individuals living in a particular situation to produce offspring that are especially suited, by their "personalities," to that situation. It still remains possible, however, that behavioral individuality is an evolutionary strategy analogous to the presumed adaptive significance of sexual reproduction itself, namely, the production of a sufficient number of (genotypically and phenotypically) diverse offspring to enhance the probability of leaving successful descendants in an ecologically and socially variable environment.

Despite the substantial theoretical and practical importance of the phenomenon (Hirsch, 1963, 1967), behavioral individuality has been largely neglected by sociobiologists. As we have seen, this is not entirely surprising, since individual variation is quite different from the sort of generalizations about age and sex classes with which sociobiology has been so successful, and with which this book is largely concerned. Nonetheless, generalizations cannot be considered successful if their success occurs at the cost of artificially leveling otherwise significant features of a species' behavioral landscape. It can therefore be hoped that eventually an evolutionary perspective on animal social behavior will go beyond homogenizing our perceptions of animals into age, sex, and other classes, and will incorporate a greater appreciation of behavioral individuality as well.

Summary and Conclusions

1. Young of the year emerge from their natal burrows at weaning; they spend progressively more time above ground with age, and within a few weeks they are foraging actively and ranging widely.

2. Young Olympic and hoary marmots are tolerated by colony residents; young yellow-bellied marmots have fewer interactions with colony

residents other than their mother; and young woodchucks interact with their mother and their littermates primarily during nursing. After weaning, young woodchucks are generally independent, and they disperse without experiencing interactions with many other individuals.

3. Play is prominent among the montane species; sexual dimorphism in play and other social behaviors develops earlier in *M. flaviventris* than in either *M. olympus* or *caligata*.

4. The early experience of marmots seems to prefigure subsequent patterns of social behavior, and may predispose individuals toward different patterns as well.

5. The recognition, identification, and interpretation of individual differences constitutes a substantial challenge for both theory and empirical study of marmot sociobiology, and for sociobiology in general.

6. Pioneering work on behavioral profiles in yellow-bellied marmots has identified distinct personality types that correlate with observed living situations. Similar individual differences appear to exist among hoary marmots as well.

7. The effect of behavioral individuality may well be substantial, especially on the process of dispersal vs. recruitment and, possibly, on interspecies differences as well.

8. Individual differences should be distinguished from differences based on differing biological or social roles.

9. As hoary (and possibly, other) marmots grow older, they appear to become more consistent in their behavioral profiles and also more unlike each other; this empirical generalization of age-related behavioral individuality may apply to other living things as well.

CHAPTER 8

Male/Male Competition

A sociobiologic view of animal behavior suggests that bodies are essentially vehicles for the propagation of genes. Male bodies differ from female bodies in one fundamental aspect—the production of sperm as opposed to eggs—and in many derivative ways, including (in many but not all cases), a tendency for larger size and greater aggressiveness. Among marmots, as among all mammals, females provide more "parental investment" (Trivers, 1972) than do males; hence, the former are a limiting resource for the reproductive success of the latter. A single male can inseminate many females, and those that do so will be strongly favored by natural selection, provided, of course, that such conceptions generate offspring that survive to reproduce. Marmots are therefore predicted to reveal signs of male/male competition, oriented especially toward ultimate reproductive success via the fertilization of females.

Dominance Relationships in M. olympus and caligata

Hoary and Olympic marmots generally experience the most unstructured social relationships of the North American species. In fact, no clear dominance structure is apparent among young of the year or yearlings. Among adult females, dominance is influenced by reproductive status and other components of individual differentiation, but it is not generally a pronounced part of marmot social life. Among the different age and sex classes, however, a dominance trend is readily discernible: adult males > adult females > two-year-olds > yearlings > young of the year. Nonetheless, the outcome of any given interaction is not rigidly predictable,

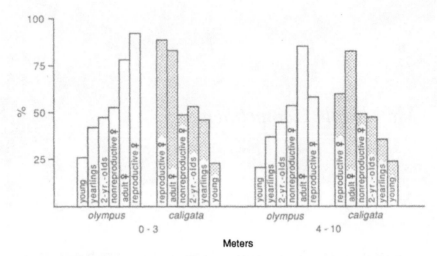

Fig. 8.1. The percentage of dyadic encounters in which various age and sex classes of Olympic and hoary marmots (*M. olympus* and *caligata*) were judged to be dominant, as a function of distance from the home burrow. Based on a minimum of 8 individuals and 78 observation-hours for each category in *olympus* and a minimum of 12 individuals and 227 observation-hours for each category in *caligata*.

and is apparently independent of any spatial referent as well. Except for reproductive females, dominance during intracolonial encounters is unaffected by proximity to home burrow in either species (Fig. 8.1). The percentage of dyadic encounters in which a given individual was judged to be dominant while 0–3 meters from its home burrow vs. 4–10 meters from its home burrow is significantly higher for reproductive females of both *M. olympus* and *caligata*[1] but unchanged for all other sex and age classes.

Instead of personal territories or even individual home ranges, a general "colony range" exists for Olympic and hoary marmots: an area of about 4–10 hectares beyond which members of the colony rarely stray but within which (with certain exceptions; see below) any colony member freely utilizes any place at any time. Very little is currently known about the social relations of the various marmot species inhabiting the USSR, although certain intriguing data have been published; for example, *M. camtschatica* (Zharov, 1972) has been reported to occupy a "stock range" of 21.8 hectares. If this is comparable to "colony range," it

[1] binomial test, P < .01

is substantially larger than that found among any of the North American species.

Among Olympic and hoary marmots, no special areas are frequented exclusively by particular individuals, and no apparent prohibitions restrict the movement of most animals. During the early morning and early afternoon "visiting periods," this unusual social plasticity is apparent: an Olympic or hoary marmot's home is certainly not his or her castle. All occupied and most unoccupied burrows are generally entered by nearly all colony residents during this time. An observer arriving after the early morning visiting period would be hard pressed to assign any animal to its home burrow, with the exception of newly emerged young of the year and their mothers.

The seasonal and individual frequencies of burrow visiting reveal that adult male Olympic marmots are the most active burrow visitors in June and July (Fig. 8.2). Yearlings' low levels in June correlate with their general inactivity at that time, also reflected in greeting frequency. Visiting by two-year-olds is quite frequent; it is possible that this restlessness of two-year-olds presages their imminent dispersal, being analogous to premigratory restlessness among birds. By August, visiting levels are comparably low for adults and yearlings, correlating with the late-season decline in greeting levels and social activity in general. The newly emerged young are the least active visitors.

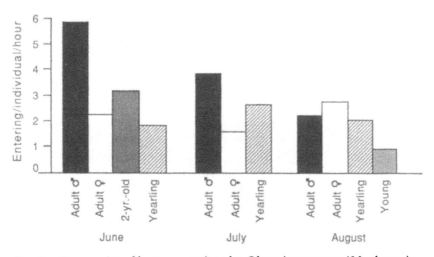

Fig. 8.2. Frequencies of burrow enterings by Olympic marmots (*M. olympus*). Based on the following observation-hours: June, 147; July, 148; August, 111.

Figs. 8.3–8.7. Play-fighting by Olympic marmots (*M. olympus*). 8.3, *top opposite*: Initiating contact while in the upright posture for play-fighting; note open mouths. 8.4, *center opposite*: Leaning against each other in early stages of upright play-fighting. 8.5, *bottom opposite*: Continuation of upright play-fighting, with extensive mouthing. 8.6, *top*: Termination of upright play-fighting, with a prolonged greeting. 8.7, *bottom*: Feeding side by side after conclusion of play-fighting. (Photos by M. D. Hutchins)

Among some species, a subordinate animal can be recognized as one that terminates an interaction. Among both Olympic and hoary marmots, however, interactions are generally so overlain with a fabric of "playfulness" as to obscure the agonistic component. Thus, apparent fights were often interrupted for no apparent reason, or simply to initiate new ones. Take the following observation of *M. olympus* in mid-July (from my field notes): "Adult female rapidly approaches yearling, who is sitting on rock. Upright fight follows immediately, with much growling and tooth-locking. They separate, still on hind legs, and pause about 3 seconds. Then, female prances away, stiff-legged, to adult male (about 3 meters) and they begin an upright fight for 10 seconds, leaving yearling standing alone. Then all three begin feeding, within about 3 meters of each other." In this case, there is no reason to believe that the adult was subordinate to the yearling, although she obviously terminated the encounter. Furthermore, the yearling made no effort to withdraw from the encounter (Figs. 8.3–8.7).

Among Olympic marmots, 38 of 56 greetings and 24 of 50 upright fights between adult males and young of the year were terminated by the adult. Only about 10% of all fights had clear outcomes.

Satellite and Colony Males in M. olympus and caligata

The most obvious cases of social dominance among male Olympic and hoary marmots occurred when more than one adult male inhabited a colony; I observed this situation closely for 7 of 38 colony-years among Olympic marmots and 10 of 74 colony-years among hoary marmots (additional brief surveys were taken at other colonies). In all cases of multiple adult males, one individual—henceforth, the "colony male"—was always clearly dominant and the other—henceforth, the "satellite male"—was clearly subordinate. Satellites always occupied a distinct burrow, removed from the other residents' (55–150 meters away). During May and June, satellites clearly avoided colony males, although they were often chased by them nonetheless, commonly 3 to 5 times per day, with a record in the case of one Olympic marmot colony/satellite pair of 8 times in a single day, and up to 3 minutes in a single chase.

In all direct interactions that I observed during May and June, the satellite ran away from the colony male. However, encounters within a burrow might be frequent, as shown by these field notes taken at an Olympic marmot colony in mid-June: "Yellowtail [a satellite] is feeding near Stumpy's burrow [a colony male]; Stumpy feeding 35 meters below. Stumpy runs over and Yellowtail runs across meadow to his home burrow. . . . [Five minutes later] Yellowtail feeding near his burrow; Stumpy runs over and Yellowtail enters burrow before Stumpy arrives. Stumpy feeding around burrow about 9 minutes before Yellowtail looks out. Stumpy sees him and runs over, growling. Y reenters burrow, followed immediately by S. Growls from inside burrow for about 5 seconds, then S emerges and wanders off. Two minutes after he has gone, Y emerges again and starts feeding."

For the first 6 weeks after seasonal emergence, satellites keep to a limited feeding area, typically within 150 meters of their home burrow, appearing to choose their position with constant reference to the colony male, maintaining as great a distance from him as possible and feeding near the main colony area only when the colony male is elsewhere. However, other colony residents (including the colony male) typically visit the satellite's burrow during the day. Olympic and hoary marmot satellite males averaged 2.6 (SD = 1.1) burrow visitings per hour during June, significantly less than would be predicted by their abundance;[2] resident males, by contrast, engaged in significantly more (6.7; SD = 2.4).[3] By August, however, burrow visiting by both satellite and colony males was 2.2 (SD = 1.0) and 2.1 (SD = 0.9), respectively.

As we have seen, colony male hoary marmots are older than satellites, and a similar pattern applies to all Olympic marmot cases in which the relative ages of satellite and colony male are known. Hoary marmot colony males also weigh somewhat more than satellites (mean of 5.8 kg vs. 5.2 kg during June).[4] As the season progresses, relations between satellite and colony males become more amicable. Thus, among Olympic marmots, I observed 0.30 chases per observation-hour during June (110 hours of observation); 0.16 from July 1 to 15 (63 hours); and no chasing from July 15 to 31 (71 hours). By the first week of August, signs of avoidance behavior generally disappear, and the two males often greet each other. Seven of 10 hoary marmot satellites changed burrow residence during August, moving in with one or more of the other colony residents. In two cases, they eventually shared a burrow with the colony male.

It is conceivable that such increased social tolerance between satellite and colony males is attributable to familiarity *per se*, rather than to the seasonal waning of the breeding season. Since satellite males may remain as satellites for several years within a colony occupied by the same colony male, it is possible to test this hypothesis by comparing the chase rates between colony and satellite males when co-occupying a colony for the first year with cases in which the colony and satellite males are interacting for the second or third year. For *M. caligata*, in a comparison of the number of chases observed per hour for 4 such "first-time" colony/satellite pairs (455 hours) with comparable data for "second-time" and "third-time" colony/satellite pairs (316 hours), no difference is apparent in over-

[2] chi-square, P < .01
[3] chi-square, P < .01
[4] Wilcoxson paired test, P < .05; n = 14 colony males and 7 satellites

all occurrence of such chases,[5] or in the regression of chase rates against time of year (i.e., in the rate of decline of such chasing).[6] Moreover, there is no trend relating colony/satellite chasing behavior to whether these dyads were first-time, second-time, or third-time.[7] (These data are based on permanently marked individuals, whose identities are definitely known across years.) It therefore seems safe to conclude, at least for *M. caligata*, that the progressive seasonal diminution in colony/satellite chasing is due to seasonal changes *per se* rather than to familiarity.

I quantified the physical spacing between the satellite and colony males at an *M. olympus* colony by recording the simultaneous locations of both on enlarged photographs of the colony and then measuring the distance between the sets of points on the photograph. Forty-eight such observations made every half-hour in mid-June were normally distributed about a mean of 29.5 meters (range = 7.6–64.0; SD = 11.5). By contrast, 53 similar random observations in mid-August revealed a mean of 14.4 meters (range = 1.8–39.9; SD = 10.5).[8] A random sample of 20 such recordings obtained for the colony male and a nonparous female in mid-August yielded a mean of 11.8 meters (range = 0.6–52.1; SD = 8.7), which is not significantly different from that obtained for the colony and satellite male during the same time. These results suggest (1) that colony/satellite male avoidance was less in August than in June, and (2) that physical spacing between these animals in August was similar to the spacing between the resident male and other adults. This easing of colony male/satellite male relations is always temporary, however—the satellite is relegated to a distinctly subordinate status the following spring.

Hoary marmot colonies containing satellites occupied meadows averaging 5.9 hectares, whereas colonies lacking a satellite averaged 3.7 hectares. Satellite male hoary marmots are more likely to occur among colonies located in large meadows than among colonies located in small ones.[9] This could be because large meadows facilitate a satellite's avoidance of the colony male: one satellite, for example, was chased substantially more often by the colony male than was any other. This satellite spent much of his time within his burrow and was gone the following year. This colony also occupied the smallest meadow of any containing a satellite (2.8 hectares). Data on chase frequencies (during June) and meadow size for 7 different colonies containing satellite males revealed a

[5] t test, P > .10
[6] P > .10
[7] Spearman rank test, P > .10
[8] t test, P < .01
[9] Mann-Whitney U test, P < .01; n = 18 colonies with satellites and 21 without

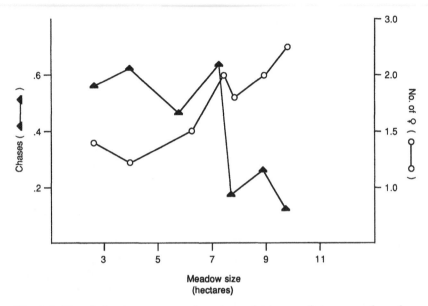

Fig. 8.8. Correlation between meadow size and (1) rate of chases per hour between colony male and satellite male and (2) number of females per colony, among hoary marmots (*M. caligata*) in Washington State. Based on observations of seven different colonies, with a minimum of 116 observation-hours at each.

significant inverse correlation: satellites at larger meadows were chased fewer times (Fig. 8.8).[10]

Colonies that offer more usable habitat could also be more attractive to otherwise dispersing males, at least partly because more space promises greater potential for male/male avoidance. In such cases, individuals might therefore be less likely to leave their natal colony (see Webb, 1981, for a comparable possibility among yellow-bellied marmots). In no case, however, was a satellite male associated with a colony containing fewer than two adult females; given that monogamous colonies constituted 33% of all those surveyed, this preferential association of satellites with polygynous sites is a significant deviation from chance,[11] suggesting that potential access to reproductive opportunities might also influence satellite occurrence. Polygyny among hoary marmots is also positively correlated with meadow size;[12] as Fig. 8.8 shows, larger meadows tend to

[10] Kendall partial correlation coefficient, $P < .01$
[11] chi-square, $P < .01$
[12] Spearman rank correlation, $P < .05$

have additional adult females as well as satellite males. Females and satellite males may well be attracted to larger meadows for the same ultimate reason—greater reproductive opportunities—although proximately females are probably influenced by meadow size as such, whereas males are probably attracted by the presence of females, as well as perhaps a greater opportunity to avoid the colony male at the same time.

There is no evidence that adult male Olympic or hoary marmots ever adopt a permanently transient life-style. By contrast, occasional transients are reported for yellow-bellied marmots (Armitage, 1974), and a strategy of "wandering" has been reported for woodchucks (Meier, 1985). It seems unlikely that wandering would be a reproductively successful strategy for males of the montane species, because of (1) the clumping of females and their relatively clear association with a resident male and (2) the severity of montane environments, which makes it essential for animals to have access to adequate burrows, and where persistent snow cover makes travel between reproductive sites particularly hazardous in the spring, when breeding occurs. It is possible, however, that the persistence of satellites among the montane species represents a tendency in this direction, such that satellites are "wanderers" who remain on the periphery of a single colony.

Male/Male Competition in M. flaviventris and monax

Among yellow-bellied marmots, satellite males—in the sense of *M. olympus* and *caligata*—are not normally found. Svendsen (1974), however, identified what he called "satellites," solitary or paired individuals that typically occupied separate areas, distinct from colony sites. I designate these individuals "peripherals" to distinguish them from the above case. About one-half of resident territorial male *M. flaviventris* are known to have spent at least one year as a peripheral before becoming a territorial resident (K. B. Armitage, personal communication). These peripherals intrude regularly upon a resident's living space and are invariably chased out. Such interactions are not frequent among yellow-bellied marmots, as they are among Olympic and hoary marmots. Peripheral females of *M. flaviventris* have virtually no interactions with colony residents, whereas peripheral females of *M. olympus* or *caligata* are unknown; all identified adult females in these species are colony residents.

In some cases, satellite male yellow-bellied marmots have been observed (K. B. Armitage, personal communication), but such arrangements are temporary, with the satellite eventually being forced to disperse. By contrast with *M. flaviventris*, both *M. olympus* and *caligata* experience a relatively high level of adult male/male chasing, especially early in the season, and satellites are common year-round residents of *M. olympus* and *caligata* colonies. The higher frequency of male/male chasing in the two high-elevation species presumably does not indicate less social tolerance than in *M. flaviventris*, but rather more, since adult male yellow-bellied marmots do not normally tolerate additional satellite males within their colony, and as a result, the frequency of male/male chasing in that species is lower than in either *M. olympus* or *caligata*.

Male territoriality is well developed among yellow-bellied marmots (Armitage, 1974): adult males often patrol their territories, increasing their visual conspicuousness by tail-flagging. This patrolling is accentuated when the meadow topography is such that the entire territory cannot be viewed from a small number of observation points. Territory size varies from 0.20 to 1.98 hectares (mean of $24 = 0.67$, $SD = 0.05$; Armitage, 1974), and it may vary from year to year with a given male. In particular, territory size tends to decrease in response to increases in the number of adjoining males, and to increase when other males are experimentally removed. In one case, removal of adults—initially the residents and then each subsequent arrival—resulted in a succession of would-be immigrants: 6 males and 2 females (Brody and Armitage, 1985). This suggests that transients may exert substantial social pressure on resident adults, not unlike that exerted by "floaters" in some passerine bird populations.

Male/male competition may well be as intense or even more so among woodchucks, although because of its asocial nature *M. monax* reveals relatively few quantifiable male/male interactions. Thus, during 994 observation-hours, I observed only 6 chases between adult males, 6 threats (tail raised, hair erected, tooth-chattering occurring 3 times), and 3 cases in which an adult male clearly moved to avoid the approach of another. In a study of central Pennsylvania woodchucks in which sex of the individuals could not be distinguished, 500 hours of field observations revealed 27 chases (only 2 resulting in an actual fight), 13 threats, and 26 avoidances, for an average of roughly one aggressive interaction per day per animal (F. H. Bronson, 1964). Rates of such interactions declined from spring to summer despite the seasonal increase in density because of the addition of dispersing young to the above-ground population in July. The average distance at which interactions were initiated was also greater during May and June (29 meters) than later in the summer (9 meters), indicating greater social intolerance earlier in the season. As F. H. Bronson (1964) emphasized, this seasonal decline in aggressiveness may be due to (1) stabilization of relationships among neighbors over time, (2) higher vegetation inhibiting visibility and, therefore, encounter frequency, and/or (3) time elapsed following the breeding season, mediated by hormonal changes (testes regress from March until October). Ultimately, these hormonal changes and their behavioral consequences are dictated by male/male competition for reproductive success.

Certain rare but important events in the lives of animals may not be adequately reflected in simple frequency measures. For example, predation, although uncommon, seems to have exerted a profound impact on marmot biology. Similarly, severe fights among woodchucks may be significant, although rare because the species is so asocial. Indeed, perhaps the species is asocial because fights, when they do occur, are potentially damaging. Up to 9% of 685 trapped Pennsylvania woodchucks revealed serious bite wounds, including tails bitten off, and adult males made up a substantial proportion of the injured (Fig. 8.9). The decline in wounding apparent from these data is paralleled by a decline in the frequency of ag-

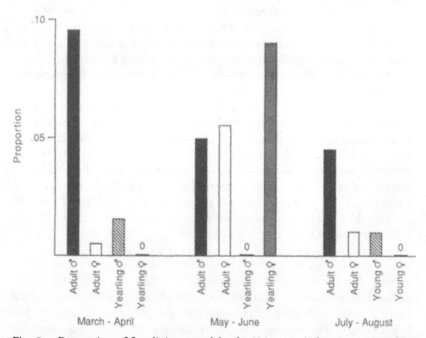

Fig. 8.9. Proportion of free-living woodchucks (*M. monax*) showing serious bite wounds. Based on 685 trapped animals and a minimum of 16 individuals in each category. (Data from F. H. Bronson, 1964)

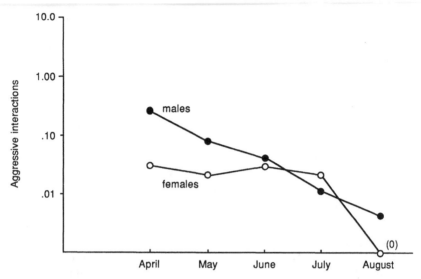

Fig. 8.10. Seasonal changes in frequency of aggressive interactions (chases, fights, and avoidances) among adult woodchucks (*M. monax*) in New York State. Data are presented per individual per observation-hour and are 4-week averages for each month, based on the following regime: April, 139; May, 177; June, 215; July, 122; and August, 118. A minimum of five individuals were observed each month.

gressive interactions involving adult male woodchucks in New York State (Fig. 8.10).[13]

A similar seasonal decline in the aggressiveness of male Olympic and hoary marmots has already been discussed. By contrast with the woodchuck case, however, I observed only 3 cases of apparent wounding out of 21 closely examined adult male Olympic marmots and only 4 cases out of 53 hoary marmots. (Such wounds, of course, could be inflicted by predators or simple accidents as well as by other marmots, but in such cases there is little reason to expect a seasonal decline.)

The receipt of aggression may predispose individual woodchucks to avoid each other. As F. H. Bronson (1963) recounts, "a yearling male that was subordinate to almost all neighbors, and which had been under considerable pressure by those neighbors, moved to a new burrow system about 1,000 feet away. At this location, it interacted with another small woodchuck (probably a yearling) that had recently moved to that burrow system from some place off the study area. The interaction was initiated

[13] Kendall partial rank coefficient, $P < .01$

at a distance of 20 feet, and after a few seconds of appraisal both animals turned and ran from each other. The interpretation was that both animals were subordinate to most of their neighbors in the areas from which they had come (true in at least the one case) and that both had been conditioned by neighbor interactions to avoid contacts with other woodchucks."

Social rank is not a useful concept among woodchucks, since relationships in such an asocial species tend to be one-to-one. "Relative rank," however, may be defined as the "relationship between the number of neighbors dominated by an individual compared to the number which dominated it," with neighbor defined as "any animal whose home range overlaps that of the individual in question" (F. H. Bronson, 1964). Within such a system, each animal establishes dominance/subordinance relationships with those animals whose home ranges overlap its own, and subordinates avoid dominants. Individuals of high relative rank—tested in the laboratory—tended to have somewhat larger home ranges in the field (F. H. Bronson, 1963). The trend is not statistically significant, however, perhaps because subordinate woodchucks occupy poorer habitat, which would in turn necessitate a larger home range, for which other adults would not compete.

It seems likely that male marmots, no less than females, are constrained by the need to acquire sufficient nourishment for normal growth as well as to accumulate adequate fat to survive hibernation. Because of the added nutritional demands on pregnant and lactating females, it seems likely that females are especially sensitive to habitats with adequate food resources and that males, in turn, are especially concerned with the defense of females. The next section examines this phenomenon among hoary marmots.

Mate Guarding and Gallivanting

Male marmots, just like males of many other species (e.g., Beecher and Beecher, 1979), can be predicted to partake of a two-part strategy: guard their female(s) against copulations with other males, which would reduce the fitness of the mated males, and seek extra copulations for themselves—even with already mated females if necessary—since additional copulations could increase their fitness if additional offspring are produced. In short, adult males can be expected to attempt precisely the behavior that they strive to prevent by other males.

Mate guarding may be defined as behavior by one member of a mated pair that directly reduces the probability that the guarded individual will copulate with another animal. Mate guarding should be distinguishable from simple sexual association in that the former should be more pronounced when the individual being guarded is likely to cuckold the guard. By contrast, if guarding is simply a consequence of mating behavior *per se*, then it should be directly correlated with sexual opportunities for the guard, rather than with the prevention of sexual opportunities for the individual being guarded. In most cases, of course, guarding and sexual association may co-occur.

Webster's *New World Dictionary* defines "gallivant" as "go about in search of adventure or excitement, especially of a sexual nature." Gallivanting, in the context of male/male competition, could well be correlated with the probability of the gallivanter's achieving additional copulations. If so, it should be appropriately combined with mate guarding if the fitness of the individuals concerned is to be maximized.

In studying mate guarding and gallivanting by male hoary marmots (Barash, 1981), I arbitrarily defined a gallivant as any excursion of more than 100 meters from the home burrow occurring during the breeding season. I paid particular attention to the behavior of males in monogamous vs. group-living situations; the latter were "colony towns," at which other males and females occupied contiguous habitat. Assuming that mate guarding and gallivanting are male reproductive strategies in hoary marmots, I made the following general predictions and tested them with data obtained during 689 observation-hours in the Washington Cascades, from 1974 through 1980 and from additional observations during 1981 and 1982.

Prediction 1. Gallivanting will be overwhelmingly a phenomenon of sexually mature males, since it is widely assumed that females benefit significantly less from outside copulations than do males; gallivanters will be repulsed by colony males, and they will seek access to females.

Test. Adult males performed 146 out of 150 observed gallivants. In all 41 cases in which gallivanters were observed after entering a new colony containing an adult male, the gallivanter was repulsed by that male. In all these cases, the gallivanter retreated within 5 seconds after encountering the colony male. This typical response of an intruder to a territorial resident is well known to ethologists, suggesting that both gallivanter and guard are abiding by "bourgeois" strategies (Maynard Smith, 1976). In any event, the immediate cost incurred by a gallivanter who encounters a colony male appears to be low, since no extensive fights resulted.

Two different colony males were live-trapped and removed from their colonies on 2 successive days each. Seven adult males were observed to gallivant into these colonies while the colony male was in captivity: in one case the gallivanter was repulsed by a nonbreeding adult female, in one case the gallivanter entered a burrow with an estrous female, in two cases the gallivanter copulated above ground with estrous females, and in the remaining three cases the gallivanter associated with the resident females, although they were lost to view before any outcome could be ascertained. Observed gallivanting rates, per animal per hour, are as follows: weeks 1–3 after emergence, 0.21 (SD = 0.17); weeks 4–6, 0.12 (SD = 0.08); weeks 7–9, 0.

Prediction 2. Mate guarding—like gallivanting—will be more prominent shortly after seasonal emergence, when females are fertile, rather than later, when they are not.

Test. Colony males were censused every 10 minutes, and their distance to the nearest adult female was recorded as < 5 meters, > 5 meters, or unknown. The results (Fig. 8.11) show that for both monogamous and group-living males, male/female distance increased significantly during the season. The average number of occasions when male/female distances were less than 5 meters declined significantly for both monogamous and group-living males from weeks 1–3 to weeks 4–6. By contrast, there was no change in the frequency of male/female distances exceeding 5 meters.

Further tests consist of recording the sex of the departing individual whenever male and female were within 5 meters and one of them walked away. The results (Fig. 8.12) show that (1) early in the season, females walked away from males significantly more often than males walked away from females, with no difference between monogamous and group-living females in this respect, and (2) later in the season, this pattern disappeared

	Less than 5 m		Greater than 5 m	
	Weeks 1-3	Weeks 4-6	Weeks 1-3	Weeks 4-6
Monogamous	0.70 (0.3) → * ← 0.36 (0.2)		1.2 (0.7) → ns ← 1.9 (0.6)	
	↓	↓	↓	↓
	**	ns	ns	ns
	↑	↑	↑	↑
Group-living	1.88 (0.4) → ** ← 0.33 (0.2)		1.0 (0.8) → ns ← 1.4 (0.3)	

Fig. 8.11. Male/female distances per male per observation-hour. Based on 613 observation-hours on 15 different males in each case, with a minimum of 40 observation-hours and three different males in each category. Data are presented as means, with SDs in parentheses; ns = no statistical significance; * = P < .05; ** = P < .01; Mann–Whitney U tests.

	Monogamous		Group-living	
	Weeks 1-3	Weeks 4-6	Weeks 1-3	Weeks 4-6
Female departed	0.47 (0.2) → ns ← 0.38 (0.3)		0.56 (0.3) → ns ← 0.41 (0.3)	
	↓	↓	↓	↓
	*	ns	**	ns
	↑	↑	↑	↑
Male departed	0.18 (0.1) → * ← 0.40 (0.2)		→ ** ← 0.53 (0.3)	

Fig. 8.12. Number of departures clearly initiated by individuals of either sex. See Fig. 8.11 for details.

	Monogamous		Group-living	
	Weeks 1-3	Weeks 4-6	Weeks 1-3	Weeks 4-6
Female followed male	0.03 (0.02) → ns ← 0.02 (0.02)		0.08 (0.01) → ns ← 0.04 (0.03)	
	↓	↓	↓	↓
	*	ns	*	ns
	↑	↑	↑	↑
Male followed female	0.29 (0.1) → * ← 0.08 (0.1)		0.55 (0.2) → * ← 0.1 (0.1)	

Fig. 8.13. Responses of both males and females to departures by the other sex. See Fig. 8.11 for details.

such that males and females did not differ in their frequency of departures. Finally, the response of males and females to departure by the other was also recorded. The results (Fig. 8.13) show that (1) early in the season, males were significantly more likely to follow departing females than females were to follow departing males, and (2) this pattern disappeared later in the season, as a result of changes in male behavior only.

Prediction 3. Mate guarding will be more prominent when the physical location of a female renders her more likely to cuckold the colony male. The likelihood of a female cuckolding her mate should increase in direct proportion to her distance from her home burrow at any given time, since this increases her proximity to neighboring males and also reduces the probability of the colony male intercepting any gallivanting neighbor male. Accordingly, mate guarding should occur in direct proportion to the female's distance from her home burrow.

Test. On the basis of 10-minute scan samples, cases in which colony male/female proximity was less than 5 meters were categorized as to the

	Weeks 1-3		Weeks 4-6	
	Female 25 m	Female 25 m	Female 25 m	Female 25 m
Monogamous	36 → * ← 60		17 → ns ← 28	
	↓ ns ↑	↓ ns ↑	↓ ns ↑	↓ ns ↑
Group-living	49 → * ← 81		11 → ns ← 16	

Fig. 8.14. Percentage of observations in which male/female distances were less than 5 meters, as a function of the female's distance from the residence burrow. Based on 613 observation-hours on 15 different males, with a minimum of 40 hours and three males in each category; ns = no statistical significance, ★ = P < .05; binomial tests for differences in proportion.

distance of the female from her residence burrow at the time: < 25 meters or > 25 meters. The results (Fig. 8.14) show that (1) during weeks 1–3, both monogamous and group-living males attend their females more closely when females are beyond 25 meters from their residence burrow than when they are within 25 meters, and (2) these differences disappear by weeks 4–6, providing further support for Prediction 2 as well.

The tendency of colony males to attend females that are distant from their residence burrows and to do so early in the season is highlighted by the fact that on eight occasions, males made conspicuous runs across the colony, to within 5 meters of the female, apparently when they became suddenly aware of the female's location. On seven of these eight occasions, the female was more than 25 meters from her residence burrow, and all eight of these events occurred during the weeks 1–3. During this time, females were seen to enter a burrow 124 times while males were above ground; males entered burrows while females were above ground only 25 times.

Prediction 4. Males will initiate and seek to maintain proximity to females more than vice versa.

Test. As we have already seen (Fig. 8.12), males depart from females less than females depart from males. Males also follow departing females more than females follow departing males (Fig. 8.13).

Prediction 5. Group-living males will mate-guard more than monogamous males, since they are at greater risk of being cuckolded.

Test. During weeks 1–3, group-living males were closer to their females than were monogamous males; this difference disappeared by weeks 4–6 (Fig. 8.11). Group-living males were also less likely to depart from females early in the season than were monogamous males (Fig. 8.12), and they followed departing females more than did monogamous males (Fig. 8.13). Surprisingly, however, group-living males are not more likely than monogamous males to guard a female who is far from her residence burrow (Fig. 8.14), although the difference is in the predicted direction. This may be because group-living males have other females which must also be guarded.

I recorded the location of the nearest resident adult female when males began 31 different gallivants. For group-living males, 4 of 17 gallivants occurred when the nearest adult female was out of her burrow. Hence, 13 of the 17 gallivants by group-living males occurred while the nearest female was in a burrow. For monogamous males, 10 of 14 gallivants occurred when the nearest female was out of her burrow, and only four occurred when she was in a burrow. This difference fails to reach statistical significance, but the trend nonetheless suggests that monogamous males are more likely than group-living males to go gallivanting when their female is out of her burrow.

Prediction 6. Males whose females are nonreproductive during a given year will mate-guard less and gallivant more whereas males whose females are reproductive during a given year will mate-guard more and gallivant less.

Test. Since hoary marmots reproduce biennially, half the females are fertile and half nonfertile during any given year. Accordingly, the benefits of gallivanting for a bigamous male should remain approximately constant from year to year. However, the costs should vary directly with the status of colony females: a male whose female is nonfertile cannot be cuckolded. Group-living males are often polygynous; therefore, at least one of their females is typically fertile during any given year. Hence, Prediction 5 cannot be tested crisply with them. However, monogamous males are affiliated with either a fertile or nonfertile female each year. The gallivanting rate, per male per observation-hour, during weeks 1–3 was 0.26 for group-living colony males, 0.19 for group-living peripheral males (those living in the interstices of colonies but not clearly satellite at any one), and 0.21 for monogamous males.[14] However, these data for monogamous males are combined, regardless of the reproductive status of their female. Data on gallivanting rates are available for nine monogamous males, each of whom was associated with a female of known re-

[14]Kruskal-Wallis one-way ANOVA, $P > .10$

productive status: four with fertile females and five with non-fertile females.

The data (Fig. 8.15) show that (1) monogamous males paired with fertile females gallivanted less during weeks 1–3 than did monogamous males paired with infertile females, and (2) this difference disappeared by weeks 4–6. Similarly, data indicative of mate guarding (spatial proximity, extracted from Fig. 8.11 and presented in Fig. 8.16), show that (1) monogamous males paired with fertile females stay closer to their females during weeks 1–3 than do similar males paired with nonfertile females, and (2) this difference disappears by weeks 4–6.

Prediction 7. Among group-living males, the frequency of gallivanting

	Weeks 1-3	Weeks 4-6
Female non-fertile	0.46 (0.3) → * ← 0.20 (0.2)	
	↓ ↓	
	* ns	
	↑ ↑	
Female fertile	0.10 (0.04) → ns ← 0.16 (0.03)	

Fig. 8.15. Gallivanting rate per animal per hour of monogamous males as a function of whether their females were fertile or nonfertile. Based on 149 observation-hours, with a minimum of 20 hours and two males in each category; ns = no statistical significance, * = P < .05; Mann-Whitney U test.

	Less than 5 m		Greater than 5 m	
	Weeks 1-3	Weeks 4-6	Weeks 1-3	Weeks 4-6
Monogamous	0.9 (0.4) → ns ← 1.1 (0.4)		1.4 (0.7) → ns ← 2.0 (0.6)	
	↓ ↓		↓ ↓	
	* ns		ns ns	
	↑ ↑		↑ ↑	
Group-living	2.6 (0.4) → * ← 0.8 (0.5)		2.0 (1.0) → * ← 1.5 (0.6)	

Fig. 8.16. Recorded male/female distances per male per observation-hour, comparing monogamous males whose females were fertile with others whose females were nonfertile. Based on 149 observation-hours, with a minimum of 20 hours and two males in each category; ns = no statistical significance, * = P < .05, ** = P < .01; Mann-Whitney U tests.

will vary inversely with the proportion of neighboring males and directly with the proportion of neighboring females.

Test. Group-living males may be characterized by the sex of their adult neighbors. I obtained an index of "adult neighbor sex ratio" by constructing a circle with radius 100 meters using the residence burrow of each focal male as the center, and recording the sex of all other adults whose residence burrows occurred within the arbitrarily defined circle. Data were obtained in this way for eight different males during various times; adult neighbor sex ratios for three males were recorded in two consecutive years, and one male's adult neighbor sex ratio was obtained for three years, yielding a total of 17 marmot-years' data. Of these, five cases yielded a male:female adult neighbor sex ratio exceeding 1.5 (high ratio of neighboring males to females), eight were between 1.5 and 0.5 (intermediate ratio) and four were less than 0.5 (a high ratio of neighboring females to males). The gallivanting rates showed significant differences, in the direction predicted; i.e., gallivanting rates varied inversely with the adult neighbor sex ratio. Thus, the *per capita* gallivanting rates were as follows: ratio > 1.5, 0.14 (SD = 0.02); ratio 1.5–0.5, 0.30 (SD = 0.02); and ratio < 0.5, 0.56 (SD = 0.1). This is based on 174 observation-hours, with a minimum of four males and 35 hours in each category.[15]

Prediction 8. At colonies containing satellite males, colony males should gallivant less and mate-guard more than at colonies lacking satellites.

Test. Two hoary marmot colonies, each containing one colony male and one satellite male, were observed during May and June, 1981 and 1982, for a total of 256 hours. Gallivanting rates for the colony males were 0.08 (SD = 0.03) during weeks 1–3 and 0.02 during weeks 4–6 (SD = 0). These rates are significantly lower than those obtained earlier for colony males lacking satellites.[16] Similarly, data on male/female distances, collected as in Fig. 8.11, averaged 2.55 (SD = 0.33) during weeks 1–3 and 0.52 (SD = 0.21) during weeks 4–6; once again, the prediction is confirmed: when satellite males are present, colony males remain closer to their females than when satellite males are not.[17]

Although mate guarding and gallivanting are best seen as an integrated competitive reproductive strategy of males, these behaviors are necessarily mutually exclusive at any one time: a guarding male cannot simultaneously be gallivanting, and vice versa. In arriving at his reproductive strategy, each male should be selected to consider the immediate

[15] Kruskal-Wallis one-way ANOVA, $P < .05$
[16] Mann-Whitney U tests, $P < .01$
[17] Mann-Whitney U tests, $P < .01$

local situation, especially the reproductive state of his female(s), the adult neighbor sex ratio, and the immediate presence of other males. With sufficient data on costs and benefits of the various alternatives, an appropriate analysis may well be suggested by game theoretic models, solving for an evolutionarily stable strategy (Maynard Smith, 1982), since the payoff to each participant seems likely to vary with what the others are doing.

Moreover, such payoffs are probably frequency dependent. Consider a hypothetical population composed only of mate-guarders: since marmots tend toward polygyny, some females would likely be unguarded. This would render the population susceptible to invasion by gallivanters. Furthermore, if all other males are guarding, then a gallivanting male would not suffer the possible cost of being cuckolded by another gallivanter. As gallivanting spread, however, it would become increasingly costly until an equilibrium was reached. This equilibrium would depend on many additional social and ecological factors, as well as the degree of gene exchange between colonies (gene exchange data are presented by Schwartz and Armitage, 1980).

Initial efforts to model evolutionarily stable systems of this sort are laudable (e.g., Rubenstein, 1980), although the necessary field data are still lacking. In particular, no data currently exist on the reproductive success of gallivanters and mate-guarders. The existence of alternative behavioral strategies suggests the existence of evolutionary stability, but this remains hypothetical unless the fitness of both strategists is shown to be equal. Although it is possible that satellite and peripheral males are following a type of pure strategy (gallivant only, since they have no females to guard), colony males are more likely engaging in a mixed, conditional strategy: guard if your female(s) are reproductive, if the adult neighbor sex ratio is high, and/or if a satellite male is present, and gallivant if your female(s) are nonreproductive and/or if there is little immediate competition from other adult males. Since female marmots cease being sexually receptive shortly after they are mated, a guarding male could, in theory, switch quickly to a gallivanting strategy after his female has been inseminated. It seems unlikely, however, that colony and noncolony males represent an evolutionarily stable strategy (or even a culturally stable strategy: see Dawkins, 1980), since reproductive success of the latter appears to be low. Convincing data on this point also are not yet available. Since, as we have seen, satellite males tend to be younger than colony males, some of the former almost certainly accede to the status of the latter, as colony males die or become moribund. Noncolonial males can therefore be seen as biding their time, awaiting promotion to colony male status, while, in turn, the colony males seek to maintain their situation, and, whenever possible, to better it at the expense of neighboring males—so long as they do not unduly compromise their own likely paternity in the process.

Finally, there are numerous currently unevaluated complications for the evolution of mate-guarding and gallivanting behavior, and for our understanding of them. For example, there may be costs due to increased liability of predation. The typical marmot response to a predator's appearance is to enter a burrow, and a gallivanter, being away from his home area, would likely be (1) less familiar with the location of refuge burrows, (2) farther from such burrows, and (3) more conspicuous to predators, especially while traversing intervening terrain such as snowfields. In addition, gallivanting males may increase the risk to predation of their females and offspring, since while gallivanting, males are not available to spot predators and give alarm calls. This might be especially important at monogamous sites, where no additional adults are available to sound an alarm. A male's confidence of paternity might therefore be expected to influence his tendencies to mate-guard or gallivant; all things being equal, those with higher confidence of paternity should be less likely to gallivant. Similarly, the behavior of colony males with satellites might reflect a lower confidence of paternity than would be the case for monogamous colony males, with group-living males somewhere in between.

The preceding discussion of mate-guarding and gallivanting was limited to hoary marmots. Among Alpine marmots, comparable gallivanting occurs between adjacent colonies, and, once again, it is most common early in the season and is performed largely by adult males (Barash, 1976a). Colony males typically perch uphill on prominent boulders 25–35 meters above their colonies. From this vantage, they detect incursions by adjacent males; they successfully intercepted all 30 attempted incursions that I witnessed. By contrast, visiting adult females and yearlings, the latter especially frequent in August, are permitted to remain in the colony area, mixing freely with the colony residents.

The situation among yellow-bellied marmots is unclear: it appears that intercolony gallivanting by adult males is less frequent than among the high-elevation species but that transients and peripherals are significantly more frequent (Svendsen, 1974; Armitage and Downhower, 1974). Adult male yellow-bellied marmots exposed to such transients or peripherals— at least early in the season—can be expected to reveal predictable patterns of increased gallivanting if the latter are female, and/or increased mate-guarding if they are male. In addition, some female yellow-bellied marmots occupy highly dispersed mini-habitats, far from other conspecifics, and some adult males, in turn, appear to wander widely rather than defending colony territories (K. B. Armitage, personal communication). These males may have adopted a behavioral strategy of "chronic gallivanting," seeking out the dispersed, noncolonial females.

Mate-guarding among woodchucks is limited to the short-lived male/female breeding associations of early spring. During that time, guarding may be intense although brief, and hence not easily observed. I once observed an adult male *M. monax* approach a burrow that was temporarily being occupied by an adult male and female, on April 3, in central New York State. The female was right by the burrow entrance; the resident male, about 5 meters away. The female entered the burrow when the strange male approached, and the resident male gave a brief "whistle" and ran toward the stranger, who immediately ran away.

Comparative Aspects of Male/Male Competition

It seems reasonable that the montane species, being more colonial, offer greater opportunities for male/male competition than are found among the relatively solitary woodchucks, since in the former cases, a single male should be more capable of monopolizing a number of females (although G. Svendsen, of Ohio University, has found a male woodchuck defending up to three females). It may also be significant that breeding seasons are progressively more extended at lower elevations: the woodchucks that I studied, for example, were fertile between mid-March and late April or early May. The longer vegetative growing season experienced by *M. monax* may well have relaxed the selection pressure for a short breeding season experienced by the montane forms. In addition, since about one-half of yearling woodchucks reproduce, and since yearlings emerge from hibernation later than adults, thereby reproducing later (de Vos and Gillespie, 1960), the potential reproductive opportunities of male woodchucks last longer than the other species'. As a result of these factors, adult male *monax*, via sequential polygyny, might experience male/male competition comparable to that generated by simultaneous polygyny in the colonial forms. As we have already seen, adult male Olympic and hoary marmots spend more time than any other age or sex class "looking out" of their burrows during June; by July, this difference has disappeared. Similarly, adult males spend more time outside of their burrows and are less subject to the midday activity lull (Fig. 8.17). While outside their burrows, adult males in May and June are also more watch-

Fig. 8.17. Dominant male hoary marmot (*M. caligata*) outside his burrow at 1230 hrs in early July, when all other colony members are below ground. (Photo by D. P. Barash)

ful than females, as shown by the extent to which they interrupt their foraging to "look up." This male/female difference suggests that "looking up" may not be motivated solely by predator avoidance. Furthermore, during July and August, the situation is reversed, and adult females "look up" more than do adult males (Fig. 8.18).[18] This pattern is consistent with the hypothesis that males are looking up for reasons other than predator alert alone; namely, male/male competition, specifically via pressures engendered by the need for mate-guarding and the opportunities for gallivanting.

Consistent with this hypothesis, monogamous male hoary marmots spend less time looking up than do group-living males (15.3 seconds/minute, SD = 6.9 vs. 24.2 seconds/minute, SD = 7.0),[19] even though predator alert should lead to the opposite finding: less looking up in group situations where there are other potentially vigilant individuals. (Of course, this need not be true if groups tend to attract more predators.)

Greater male watchfulness seems to exact a cost, as shown by the fact

[18] t tests, P < .05
[19] Mann-Whitney U test, P < .01

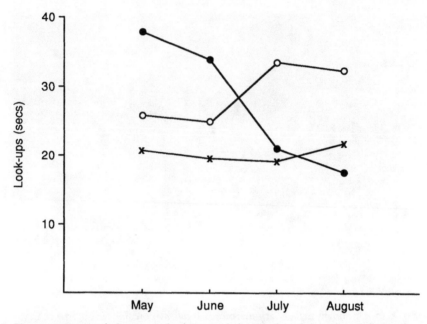

Fig. 8.18. Seasonal changes in looking-up behavior for adult male, adult female, and yearling hoary marmots (*M. caligata*). Data are presented as number of seconds looking up per individual per minute, and are based on a minimum of eight different individuals in each cohort and a minimum of 3 observation-hours of each individual during each month.

that male Olympic marmots during July spend more time foraging than they do in June, whereas foraging rates for adult females and yearlings are similar these 2 months. This finding is also consistent with the hypothesis that intense male social activity early in the season restricts their foraging to a level below that of the adult females and yearlings, which are not party to the rigors of male/male competition.

I calculated "relative daily weight gain" for adult male hoary marmots as follows: change in weight between successive weighings, divided by the time elapsed, multiplied by the average weight of the individual during the measurement period. Four colony males with satellites, each weighed regularly during two successive years, averaged a relative daily weight gain of 1.2 grams per kg-day during May and June; by contrast, four colony males without satellites averaged 1.7 grams per kg-day, suggesting that sharing one's colony with satellites may be costly.[20] This is

[20] Mann-Whitney U test, P < .05

presumably due to the metabolic expense of increased chasing, the likely costs of interrupting oneself while foraging, and possibly the direct consequences of forage competition as well (although this last possibility seems unlikely: see Kilgore and Armitage, 1978).

Among woodchucks, adult males emerge about one month earlier than females, and they lose weight precipitously. As a result, newly emerged females weigh almost as much as the males, despite the fact that the latter were significantly heavier at the onset of hibernation (Snyder, Davis, and Christian, 1961). Adult male woodchucks lose 20% of their weight during the postemergence period of food scarcity, whereas by contrast adult females lose only 6%, (nonreproductive) yearling males 8.1%, and yearling females 5.8% (Snyder, Davis, and Christian, 1961). Significantly, adult males don't begin regaining weight until mid-April, 2 weeks later than male yearlings, which do not engage in comparable male/male competition. There is no reason to think that this male/female difference in postemergence weight loss is due to male/female differences in metabolic efficiency, since if anything males seem to metabolize more efficiently, at least during hibernation: adult males lost an average of 15% of their body weight during the winter, whereas adult females lost an average of 24%.

Male/male aggressive competition is especially intense just after spring emergence, so that weight loss by adult males appears to be due to the metabolic demands of competition itself, as well as the fact that less forage is available for early emergers (all of whom are males). Finally, because the metabolic rates of endotherms during hibernation are between 1/30 and 1/100 of resting levels, early emergers are subject to much greater metabolic stress than those remaining torpid. It seems likely that the demands of male/male competition necessitate this additional metabolic imposition. Breeding occurs shortly after the females emerge, by which time the adult males have already been out for about one month and, presumably, have solidified the dominance/subordinance relationships that lead to reproductive success.

This proposition, although highly likely, is also difficult to prove: I trapped four adult male woodchucks and introduced them into new areas of good woodchuck habitat in early April in New York State. Not one established himself; by contrast, two of three adult woodchucks released in different areas during late June set up residence. These findings are consistent with the hypothesis that early spring social maneuvering, despite its attendant costs, is important for success in male/male competition, and, ultimately, reproductive success. (Admittedly, however, they are not inconsistent with other interpretations.)

As summer proceeds, the territories of adult male woodchucks remain unchanged, whereas the home ranges of adult females increase in size. This has been interpreted as showing that males are defending hibernacula, rather than females (Meier, 1985). However, if males are in fact defending females, there would be no reason for them to expand their territories during the summer, since females are not sexually receptive at that time; defense would be crucial only during the spring postemergence period, when females are sexually receptive. The fact that male woodchucks do not expand their territories during the summer could be consistent with defense of females and/or hibernacula.

Subordinate woodchucks appear to be at a competitive disadvantage with respect to dominant individuals in regard to foraging as well as the insemination of females. Subordinate individuals spend a lower proportion of time foraging and a higher proportion "down-alert," crouching and looking out for their more dominant neighbors. Moreover, low-ranking individuals showed especially high down-alert frequencies in April and May, when aggressiveness was high (F. H. Bronson, 1963). And finally, a weak but positive correlation exists between adrenal weight and proportion of activity spent in subordinance postures such as "down-alert" (Christian, 1970), suggesting a stress-related effect.

Variability seems to be the rule among yellow-bellied marmots, yet certain trends are apparent that are consistent with male/male competition. Thus, male yearlings are more aggressive than female yearlings, greeting somewhat less and chasing, being chased, and avoiding more (Armitage, 1973, 1974). Moreover, adult males initiate both absolutely and proportionately more agonistic encounters with yearling males than with yearling females (Downhower and Armitage, 1981). Yearling yellow-bellied marmots have attained their age of dispersal; by contrast, yearling hoary and Olympic marmots are still one year away from dispersal; not surprisingly, male/female differences are less clearly developed among them. For example, the ratio of yearling male/female greetings to male/male greetings is 4.77 for *M. flaviventris* (Armitage, 1974), and 1.23 for *M. caligata*. The former is significantly larger than the latter,[21] and the latter is not significantly different from 1. Similarly, adult male yellow-bellied marmots were seen to chase female yearlings 18 times and male yearlings 60 times (Armitage, 1974), for a ratio of 0.30, whereas among hoary marmots, this ratio is 0.89, which again is significantly larger,[22] and not significantly different from 1.

[21] binomial test, P < .01
[22] binomial test, P < .01

In *M. flaviventris*, as in *M. caligata*, older males seem more likely to have a harem (Armitage, 1974), and the replacement of one colony male by another is relatively frequent among yellow-bellied marmots, apparently a result of male/male competition. The average tenure for *M. flaviventris* colony males is 2.4 years, and turnovers are more frequent among larger harems (Armitage and Downhower, 1974). Among hoary marmots, by contrast, I have observed 27 turnovers of males, yielding an average tenure of 3.2 years (SD = 0.5).

Although it is tempting to conclude that male/male competition is less in *M. caligata* than in *M. flaviventris* (and, presumably, less yet in *M. monax*), this seems premature. Longer tenure length may simply reflect longer maturation time—although, admittedly, maturation time could legitimately be considered one of those factors that contributes proximately to the degree of male/male competition experienced by any species. On the other hand, since degree of polygyny seems likely to correlate directly with intensity of male/male competition, the montane species presumably experience more male/male competition than does *M. monax*, since the former are almost certainly more polygynous. *M. flaviventris*, *olympus*, and *caligata* appear to be about equally polygynous: i.e., their male:female breeding ratios seem to be roughly comparable. Moreover, as we shall see, sexual dimorphism is significantly lower among woodchucks than among the montane marmots, consistent with an anticipated lower level of male/male competition among the former.

Among *M. caligata* colony males, the average tenure for 12 bigamists was 2.5 years, whereas for nine monogamists it was 3.7.[23] This may be because bigamous males are subject to greater competition from immigrants, because their larger harems make their colonies a greater prize for a would-be colony male. Or the tenure of bigamous male hoary marmots may be shorter because such individuals are subjected to more intracolony competition, since bigamous males produce nearly twice as many two-year-old males per year as do monogamists. Unlike yellow-bellied marmots, male hoary marmots are sometimes replaced by maturing males from within their own colony (see Chapter 13). This produces an interesting situation in which the tenure of successful bigamous males is threatened by their own success, in the form of their own maturing offspring.

On the basis of a total of 22 litter-years, monogamous male hoary marmots averaged 4.3 (SD = 0.2) young per litter and bigamous males averaged 4.4 (SD = 0.2). The former averaged 0.39 litter per female, for a

[23] Mann-Whitney U test, P < .01

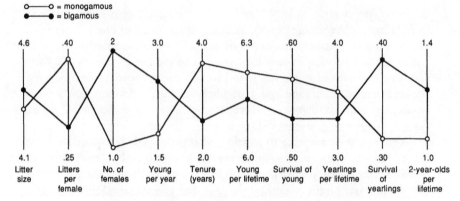

Fig. 8.19. Progression of factors leading to the lifetime reproductive success of monogamous and bigamous hoary marmots (*M. caligata*) in Washington State. The factors are multiplicative; e.g., for monogamous males, average litter size = 4.3 × 0.39 litter per female × 1 female per year = 1.68 per year × 3.7 years tenure = 6.22 young per lifetime, etc.

total of 1.68 young per monogamous male per year; the latter averaged significantly fewer, 0.28 litter per female,[24] which when multiplied by 2 gives a total of 2.46 young per bigamous male per year. Bigamous males, accordingly, appear substantially more fit than do monogamists. However, when these figures are multiplied by the average tenure for monogamous and bigamous males (3.7 and 2.5 years, respectively) they become 6.22 for monogamists and 6.15 for bigamists—essentially identical.

Thus far, these findings suggest that bigamous and monogamous males are equally fit. However, mortality among hibernating young and yearlings is somewhat higher among monogamous than among bigamous colonies (Fig. 8.19), so that bigamous males produce somewhat more emerging two-year-olds. Because of the high overwinter mortality of hibernating young (see Chapter 11), reproductive success is better measured by the number of successful dispersers produced, rather than by the number of young of the year. A similar pattern relating polygyny to reproductive success has been reported for yellow-bellied marmots (Downhower and Armitage, 1971), although without respect to adjustments due to tenure length. Thus, male yellow-bellied marmots with two or more females are more fit than monogamists, with male reproductive success increasing almost linearly with increase in his harem size (Armitage, 1986b). This relationship obtains despite the fact that

[24] t test, P < .05

proportionately fewer females reproduce in larger harems. Males of both *M. caligata* and *flaviventris* should be selected to upgrade their reproductive status from monogamy to bigamy whenever possible.

Adult male woodchucks, in a sense, are all gallivanters; presumably they mate-guard as well during their brief consortship with an adult female early in the spring. They also seek to upgrade their breeding, by establishing sequential consortships. Their success in doing so, however, may well be limited by the dispersion of females. Adult male yellow-bellied marmots, for their part, do not actively recruit additional females into their colonies (Downhower and Armitage, 1971), and their frequency of gallivanting appears to be low. As with the other montane species, yellow-bellied marmots defend their harems. They also tend to behave less aggressively toward yearling females than toward males.

Just as adult male yellow-bellied marmots do not appear to recruit yearlings into their harems actively, adult male hoary marmots do not overtly solicit the addition of two-year-olds. On two occasions I witnessed an apparently dispersing two-year-old female enter a monogamous colony: both times the male behaved with evident interest and agitation, but in neither case did the female stay and in neither case did the adult male behave in any way that seemed to interfere with her departure. However, several times I have seen colony males conspicuously herd their reproductive adult females away from the area inhabited by a satellite male.

Summary and Conclusions

1. Dominance relationships are flexible among Olympic and hoary marmots, except for relations between adult males; satellite males are always subordinate to colony males.

2. Satellites tend to be younger and smaller than colony males; the former avoid the latter early in the season.

3. Satellites are more frequent when meadows are larger and when more than one female is present in a colony.

4. By contrast with Olympic and hoary marmots, yellow-bellied marmots are more consistently territorial, with males defending harems.

5. Woodchucks are the most aggressive of the well-studied marmots, but the low frequency of observed interactions among them makes generalizations difficult.

6. A syndrome of mate-guarding and gallivanting characterizes the behavior of male hoary marmots, as they seek to prevent copulations by

"their" females with other males while also seeking additional copulations for themselves.

7. The cost of male/male competition is suggested by differential patterns of weight loss and activity budgets.

8. Male/male competitive patterns develop progressively later in those species that mature later.

9. Much as it is tempting to generalize about the comparative intensity of male/male competition among different species, simple cross-species comparisons are likely to be misleading.

10. Bigamous males, and those harboring a satellite, seem exposed to a higher level of male/male competition than do monogamous males, but they also appear to experience slightly higher reproductive success.

Female/Female Competition

S ociobiology has achieved substantial insights into male/male competition and mate selection by exploring the behavioral consequences
of maleness—i.e., sperm production—and femaleness—i.e., egg production (Wilson, 1975; Barash, 1982). In particular, researchers have emphasized that males tend to be relatively aggressive sexual advertisers
whereas females tend to be coy comparison shoppers (Williams, 1966)
and that these differences correlate predictably with differences in parental investment (Trivers, 1972). Female/female competition, by contrast,
has received considerably less attention, perhaps in part because of some
male bias by research scientists, as well as the fact that female/female
competition tends to be generally subtle and less conspicuous than its
male counterpart. However, female/female competition is no less real
(see Wasser, 1983, for a sample of recent attention to this important and
often neglected issue). It seems likely that future students of marmot
sociobiology will devote more of their attention to female/female competition than has been the case thus far; emerging trends suggest that
female marmots are highly competitive in their own way.

Competitive Aspects of Harem Size

We have already seen that monogamously mated female hoary marmots produce more young of the year than do bigamously mated individuals because the latter are more likely to skip an additional reproductive year than the former. A similar pattern has been reported
for yellow-bellied marmots as well (Downhower and Armitage, 1971;
Figs. 9.1 and 9.2). It should be noted, however, that in *M. flaviventris* the

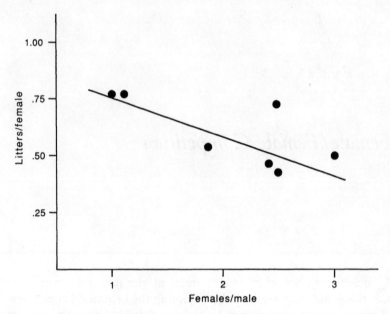

Fig. 9.1. Correlation of number of litters per female with harem size among yellow-bellied marmots (*M. flaviventris*). Five data points represent 7-year averages; two data points represent 5-year averages. (Based on data from Downhower and Armitage, 1971)

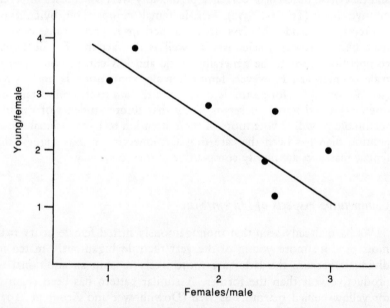

Fig. 9.2. Correlation of number of young per female with harem size among yellow-bellied marmots (*M. flaviventris*). Data as for Fig. 9.1.

number of resident female offspring does not influence the mean litter size or the mean number of litters produced per female; reproductive skipping is more frequent, on the other hand, with increases in harem size. Just as with adult males, determination of female fitness should also take account of tenure within a colony of a given size, since the ultimate evolutionary prize is connected to reproductive success over a lifetime rather than during the course of a single year. It is at least possible, therefore, that adult females actually enhance each others' lifetime fitness, as opposed to yearly fitness, for example, by providing greater potential predator avoidance, which in turn results in longer lifespan and the opportunity for greater lifetime reproductive success (see Elliott, 1975, for a model based on this assumption).

Evidence from *M. caligata*, however, does not support this view. Mortality data are available for 35 adult females, of which 22 were bigamous and 14 were monogamous. Age at death for the bigamous females averaged 6.5 years (SD = 0.3) and for the monogamous females, 6.4 (SD = 0.3); the differences are not significant.[1] In *M. olympus, caligata,* and *flaviventris,* females tend to establish lineages of genetic relatives—notably their female offspring—termed matrilines (Armitage, 1984). There is no effect of harem size on survivorship of adult female *M. caligata* or *olympus* (my data) or *flaviventris* (Armitage, 1986a). Similarly, there is no relationship between lifetime reproductive success, measured as production of young, and harem size for *flaviventris.* Such analyses, however, are particularly difficult to undertake, since a given female may be part of a matriline of different sizes for different numbers of years, alone for one year, one of two individuals for two more years, then one of three for several more years, etc.

Females of the montane marmots apparently enjoy higher reproductive success in smaller harems, because they produce more offspring. This leads to the prediction that females should respond aggressively to other females who seek admittance to their harem—if such behavior in itself is not too costly.

Field results bearing on this proposition are equivocal but suggestive. I have observed dispersing two-year-old female hoary marmots enter existing colonies on two occasions; both times they were intercepted and chased away by the resident female. In addition, I have observed 87 interactions between resident hoary marmots of adjacent colonies. Of these, 24 involved incursions by adult females; in 17 of these cases, the nonresident was promptly repulsed by a colony resident—15 of these 17 times the repelling individual was a resident adult female (the other two times it

[1] t test, P > .10

was a two-year-old female). We can further predict that the marginal fitness decrement to a resident female of adding one more female to her
harem should decline with increasing harem size: in a large harem, adding
one more female would probably decrease the reproductive success of
each constituent female less than would the transition from monogamy
to bigamy. Hence, it may be significant that monogamous females were
responsible for 11 of the observed 15 female/female repulsions. (It is also
possible that monogamous colonies, since they occupy somewhat smaller
meadows, simply provide greater opportunities for visual scanning by
the residents.)

Not surprisingly, pregnancy itself increases the aggressiveness of adult
females. The ratio of chases initiated to chases received provides a convenient index of aggressiveness and social dominance: ratios of 1 indicate
intermediate aggressiveness and status, ratios less than 1 indicate subordinance, and ratios greater than 1 indicate dominance and aggressiveness. I
calculated such ratios for 28 different adult female hoary marmots, from a
total of 60 marmot-years' data and 321 chases during May, June, and
early July; reproductive females averaged chase ratios of 1.86 (SD = 0.5),
whereas nonreproductive females averaged 1.37 (SD = 0.5)—the difference is significant.[2] The enhanced aggressiveness of reproductive females
suggests that female aggressiveness has some role in achieving reproductive success, presumably through females' enhanced success in defending
their young (from possible infanticide?) or resources around the natal
burrow. However, it does not necessarily speak to female/female competition *per se*, since aggressive pregnant and lactating females are also
aggressive toward males, and, on occasion, subadult females as well.

Among yellow-bellied marmots, recruitment of new adult females
and of yearling females varies with the number of returning adult females
(females which had been resident in the colony the previous year): with
fewer returning adult females, recruitment is higher, suggesting that the
presence of adult females somehow inhibits females from joining a harem
(Downhower and Armitage, 1971). As we shall soon see, genetic relatedness appears to influence this recruitment pattern, especially in *M. flaviventris*. Granted that some degree of female/female intolerance ultimately
contributes to restricting harem size below what it would otherwise be,
we can also inquire into the likely proximate factors that generate this
intolerance. Since, as we have just seen, pregnant females tend to be aggressive and intolerant, especially in the vicinity of their littering burrow,
the dynamics of female fertility may well be involved.

[2] t test, P < .01

Here is one possible scenario: when annual spring snowmelt is late—and, hence, the growing season for the coming summer is shortened—fewer female yellow-bellied marmots become pregnant (Downhower and Armitage, 1971). Female/female aggression is therefore reduced, and harem size accordingly tends to increase. When the growing season is long (earlier spring snowmelt), more females become pregnant, female/female aggression increases, and harem size tends to decrease. Significantly, however, this process appears to be strongly mediated by kinship: the yearlings of animals related by less than $r = .5$ simply are not accepted by adults, regardless of reproductive condition or population density (K. B. Armitage, personal communication).

If this system operates, it could have a self-damping effect on average colony size: larger harems would likely contain a larger number of pregnant and, hence, especially intolerant females, which in turn would constrain further harem growth. Smaller harems, by contrast, would be more susceptible to the addition of further females, since they would contain fewer pregnant, feisty females. On the other hand, the female inhabitants of larger colonies are somewhat more likely to skip a reproductive year, which itself reduces the tendency of larger colonies to contain more pregnant females. And if smaller colonies can be more easily monitored by their residents, then would-be immigrants could be more readily excluded by the resident females. More data are clearly needed at this point, since, for example, it is also possible that a female attempting to enter a larger colony would also have to confront a larger number of potential repellers.

Although most marmots typically hibernate in one or a small number of burrows, even in the colonial montane species the pregnant females generally spread out each spring, occupying their own burrows and behaving, in a sense, more like monogamous individuals. Among hoary and Olympic marmots, the spatial separation of parous females does not result in any lesser integration into their colonies; individuals nonetheless greet and interact every day, typically many times. Among yellow-bellied marmots, however, home ranges are often readjusted after the mating season, such that the home ranges of reproductive females frequently do not overlap (Armitage, 1986b).

Causes of Competition

The compelling evidence for female/female competition leads to a question: what are they competing for? What is the limiting resource that induces females at larger harems to reproduce less often? It appears to be a useful working hypothesis that among males of most species, competition centers around access to reproductive opportunities (i.e., females), whereas among females, competition centers around the production of successful young (i.e., resources). As a corollary, when males compete directly for resources, they are actually competing for females, since females are attracted to appropriate resources.

In *M. flaviventris* at least, food availability appears to limit the reproductive success of adult females. Thus, the date of the last snowfall (later = shorter growing season) correlates with the number of young produced by each female (Downhower and Armitage, 1971). Yellow-bellied marmot meadows in the Colorado Rockies experience a summer-long increase in standing primary production (defined as the mean dry weight of the standing crop): from 215 grams/square meter (4,745 kilojoules) on June 6 to 348 grams/square meter (11,727 kilojoules) on July 16 to 458 grams/square meter on August 9. By August, however, the available energy content of the meadows has actually decreased slightly, to 11,567 kilojoules/square meter (Kilgore and Armitage, 1978). This bespeaks the late-summer desiccation that typically affects alpine and subalpine meadows. Although, according to Kilgore and Armitage (1978), *M. flaviventris* consumes only 0.8% to 3.1% of the above-ground primary production, this may nonetheless represent a critical and limiting consumption, if it includes a large proportion of preferred plants among those available. In addition, individual females may well be food-limited in their reproductive success, since food availability may be crucial and restrictive at certain times, such as during early lactation and during late summer when the young of the year must put on sufficient weight to survive hibernation. Such constraints may well limit a female's fitness because of localized scarcity at critical times for her reproductive success.

Among yellow-bellied marmots, a significant positive relationship has been described between the actual number of emerging young and the potential number of young a female could produce, estimated from availability, biomass, and caloric value of food located within a given adult female's home range (Andersen, Armitage, and Hoffmann, 1976). A colony of *M. flaviventris* studied in Wyoming had a long snow-free growing season and abundant food. It also had an exceptionally large harem, varying from 7 to 16 adult females (Armitage, 1962). Olympic marmots consume about 17% of the available primary production in their meadows (Wood, 1973), and available food is actually much less than net production. For example, competing herbivores (deer, grouse, chipmunks, etc.) consume another 13%; moreover, because of social, thermal, and antipredator constraints, marmots almost certainly do not have the unused portion simply available for consumption whenever and wherever they wish. Weight gain is greater in animals occupying more complex mosaic plant communities than in those occupying simple habitats dominated by only two major plant species (Wood, 1973). It should be noted that the possible role of other factors, such as micronutrients or even salt, remains to be evaluated.

Female/female competition may also be especially intense during years of late snowmelt, when nutrients needed for gestation and lactation are in short supply. During the springs of 1974 and 1982, snowfall in the Washington Cascades was particularly heavy, and many hoary marmot colonies did not become snow-free until early July those years. Of nine colonies during those two years, five were bigamous and four were monogamous; reproduction at all of the latter sites occurred as expected, in synchrony with a biennial breeding schedule. By contrast, four of the five females at the bigamous sites skipped an additional reproduction those years, suggesting that under conditions of environmental stress, females that cohabit with other females may be especially susceptible to reproductive inhibition.

Given the presumed role of competition for forage and nutrients among hoary marmots, however, it is surprising that no statistical correlation can be demonstrated between litter size and growth rates.

Smaller harems do not invariably lead to higher fitness per female marmot, but the exceptions are sufficiently exceptional to be informative as well. Thus, peripheral yellow-bellied marmots have somewhat larger litters than do colony residents, but the survivorship of their young is substantially reduced (Svendsen, 1974; Armitage and Downhower, 1974); hence, they have much lower fitnesses. Weight gain is higher among colony young than among peripheral young, but not significantly so. Peripheral females also appear to be more socially stressed than their colony counterparts; the former spend more of their feeding time in "alert" postures (Svendsen, 1974). It should be emphasized that such females, although monogamous in a sense, clearly occupy lower-quality habitats and cannot be considered "successful" in excluding other females. Indeed, their tenure at such habitats is typically brief (Downhower and Armitage, 1971), and in two cases adult females even abandoned their young several weeks after the latter emerged (Svendsen, 1974).

Among a monogamous Alaskan population of hoary marmots, standing crop biomass averaged only about 100–150 grams/square meter (Holmes, 1984a), which may well have prevented more than one female from occupying a given site, thereby enforcing monogamy on the males.

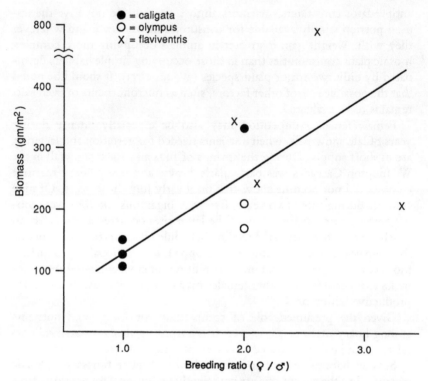

Fig. 9.3. Relationship between sex ratios of breeding adults in three marmot species and mean dry weight of above-ground vegetation at the height of the growing season. (Data from Holmes, 1984b)

Moreover, a positive correlation relates standing crop biomass and female:male breeding ratios for several marmot populations and species (Fig. 9.3).[3] In addition to food, it seems likely that female marmots compete for other resources, notably hibernacula (Holmes, 1979), and perhaps access to good-quality males as well.

Reproductive Suppression

Thus far, we have treated female/female social competition as a *result* of the negative impact that females evidently have on each others' reproductive success, with this impact itself resulting most likely from

[3] Spearman rank correlation coefficient, P < .05

food limitation. Although forage competition may well be the ultimate limiting factor in this system, female/female competition in itself can also be viewed as a possible direct *cause* of lowered female fitness in larger harems. Thus, there is a growing body of literature (reviewed by Wasser and Barash, 1983) showing that suppression of reproduction is a common strategy of female mammals. In particular, when present conditions for reproduction appear poor, but prospects for improvements in the future appear good, then females can be selected to suppress their own reproduction, essentially biding their time until conditions improve. Given the existence of reproductive self-inhibition, individuals can also be expected to suppress the reproduction of others when by doing so the suppressor increases her own chances of reproducing successfully. Such a system lends itself to numerous predictions, which have been tested by field data from hoary marmots (Wasser and Barash, 1983). These results depend especially on the phenomenon of reproductive skipping, in which adult females defer their reproduction for two consecutive years, i.e., one year beyond the pattern of alternate-year reproduction characteristic of the species.

Prediction 1. Reproductive skipping should be more pronounced among bigamously mated females, each of which must compete with another female, than among monogamously mated females.

Test. Out of 79 typical biennial breedings, I recorded nine cases of different females skipping two consecutive years. Of these nine exceptional cases, eight were bigamously mated; only one was monogamously mated. Of the 79 biennial breedings, 56 were of bigamously mated females and 23 were of monogamously mated females. Bigamously mated females are significantly more likely than monogamously mated females to skip an additional year,[4] thereby supporting prediction 1.

An alternative hypothesis, also consistent with these findings, is that bigamous females are more likely to skip than monogamous females because the former tend to occupy habitats that are lower in quality. As we saw in the previous section, the data suggest the opposite trend: higher-quality habitats (better forage) tend to be occupied by polygynous rather than monogamous colonies.

Another alternative hypothesis for the above findings is that female reproductive skipping is influenced by age and that monogamous females are a biased sample of the population's age distribution, either significantly older or younger than the mean. This hypothesis can be tested by combining available data for 1978–82; the mean age of all adult female

[4] Fisher's exact test, $P < .05$

marmots (when such age is known), is 4.8 years (SD = 1.3; n = 31). For the bigamous females it is 4.6 years (SD = 0.9; n = 16), for the monogamous females, 5.0 years (SD = 1.3; n = 9). The difference is not significant, perhaps because of the small numbers of females in each class, but it seems unlikely that differences in reproductive skipping are due to differences in the ages of monogamous and bigamous females. On the other hand, of the 12 females that skipped an additional reproductive year, the age at skipping is known for eight: their mean age is 4.3 (SD = 0.7) years, which is lower than the mean of the population as a whole, suggesting that younger females are in fact more likely to skip a reproductive year. (It is also possible that skipping females live longer and produce as many or more successful offspring over their longer lifetimes than do nonskipping females. Sufficient data are not yet available to evaluate this alternative.)

Prediction 2. Among bigamously mated females, reproductive skipping should be more frequent when both females would otherwise be reproducing synchronously than in cases in which the two females are already asynchronous. By delaying her reproduction in the former case, a female reduces competition with the other, reproductive female. This assumes, of course, that the competitive environment experienced by a reproductive female in the presence of another reproductive female—and eventually her young—is more strenuous than one faced by a reproductive female in the presence of a nonreproductive female and her yearlings.

Test. In seven of the eight cases of a bigamously mated female skipping an additional year, the nonskipping female reproduced during that year. In only one case was the skipping female associated with another female who was in the nonbreeding year of her cycle. Thus, skipping females are significantly more likely to be individuals who would otherwise be synchronous with their fellow female.[5] Moreover, during the six years for which data are available, bigamous female hoary marmots were more likely to be reproductively asynchronous than to be synchronous: there were 38 asynchronous breedings as opposed to 18 synchronous breedings.[6] This disparity in itself suggests that female reproduction is modified so as to reduce competition between the offspring of different females by reducing the likelihood that both females reproduce during the same year.

Data unfortunately are not available on the establishment of bigamous colonies. Therefore, it cannot currently be determined whether the tendency of bigamous colonies to be occupied by asynchronous females is

[5] Fisher's exact test, P < .05
[6] chi-square, P < .05

due to (1) the reproductive suppression of a newcomer or nondispersing subadult by the prior resident, (2) a greater propensity for nonbreeding females to be accepted by the breeding resident or to attempt to enter an existing colony, (3) reproductive suppression of one of two founding females by the other, and/or (4) a tendency for asynchronized females to disperse together. In any event, it seems evident that reproductive suppression—whether self-suppression or induced by another—is correlated with reproduction by the co-resident female.

Prediction 3. When reproductive skipping beyond the biennial norm occurs, it should be more frequent among subordinate than among dominant females, because given the existence of significant female/female (and/or offspring/offspring) competition, subordinates would be more likely than dominants to maximize their fitness by delaying their reproduction an additional year to avoid this competition. Moreover, if dominant females are able to suppress the reproductive success of subordinate females that attempt to breed, this could further reduce the fitness benefit to subordinates who are sensitive to suppression effects and delay their reproduction for an additional year.

Test. The seven bigamously mated females who skipped an additional year should have tended to be socially subordinate to their nonskipping co-resident female. Data on chases initiated and received during 247 hours of observation are available for four colonies that contained skippers. In three of the four colonies, the ratio of chases initiated to chases received was significantly lower for the skipper than for her nonskipping co-resident female. And whereas the fourth did not reach statistical significance, the trend was in the predicted direction (Fig. 9.4).

Since pregnant marmots tend to be more aggressive than nonpregnant marmots, however, the above data do not in themselves speak to the question of whether the low social status of skipping individuals is a cause or an effect of their failure to reproduce. Accordingly, skipping and nonskipping co-resident females were compared for their ratio of chases during the year *before* the observed reproductive skip, i.e., when both females were nonreproductive. Such data are available for three of the four cases; each of the three females who skipped an additional reproductive year was significantly subordinate to the other female during the previous year as well (Fig. 9.4).

Prediction 4. The occurrence of synchronous vs. asynchronous breeding, even without skipping, should correlate with disparity in social dominance between the co-resident females, so that synchronous breeders should be closer in rank than asynchronous breeders. If there is a wide disparity in rank, the "decision" as to whether an individual ought to

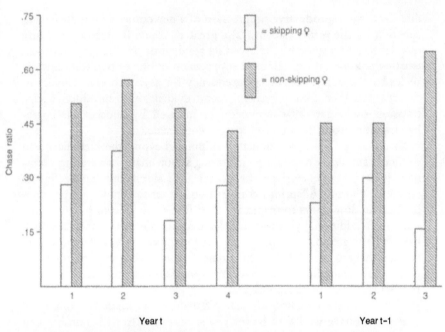

Fig. 9.4. Chase ratios for bigamously mated hoary marmot (*M. caligata*) females that skipped and that did not skip reproduction, at four different colonies. The data present the proportion of chases initiated per individual per 10 observation-hours, based on 247 hours. Data for the same individuals are also presented for the previous year for three colonies; based on 183 hours.

skip, and which individual that ought to be, should be comparatively clear. By contrast, when two prospective breeders are comparable, and each stands to benefit from breeding, we can predict a greater tendency for them to breed concurrently. Mathematical game theory seems relevant here: when the two females are asymmetric in status, reproduction may be a game of "chicken," such that if the dominant breeds, it pays for the subordinate to refrain from breeding. On the other hand, when the two females are symmetric in status, the "game" may approximate a Prisoner's Dilemma, such that each female is constrained to breed regardless of what the other one does. In this case the dilemma may not be very great, since if both females are close enough in rank and if resources are sufficient to support simultaneous reproduction by both—albeit perhaps at a lower success rate than if the other was not present—then the payoff for breeding could still be higher for each individual than it would be if one skipped.

Test. Data are available on the ratio of chases initiated to total chases (initiated plus received) for females at seven synchronous colonies and at eight asynchronous colonies. Since reproduction influences dominance status, these comparisons used the dominance status of asynchronously breeding females calculated when she was breeding. Among the seven synchronous colonies, the two females within each colony differed significantly in chase ratios[7] in only one case. By contrast, among the eight asynchronous colonies, the two females differed significantly in chase ratios six times.[8] Thus, asynchronous females appear significantly more likely to differ in dominance than are synchronous females,[9] and this difference is in the expected direction.

Prediction 5. Aside from reproductive skipping, subordinate females should express their subordinance by bearing smaller litters.

Test. Unlike primates, for example, among which singleton births are the rule, marmots produce relatively large litters, and, hence, reproductive suppression can be manifest by reduction in litter size, as well as by reproductive skipping. Adult females were identified as clearly subordinate to their co-resident adult female when their ratio of chases initiated to chases received was less than 0.5; similarly, clearly dominant individuals had a chase ratio greater than 2. Data are available for the number of young produced by 12 clearly subordinate females and by 14 clearly dominant females (Fig. 9.5). Subordinate females produced significantly fewer young than did dominant females, and this difference persisted in subsequent years, when their offspring became yearlings and, eventually, two-year-olds.

It is also possible, of course, that dominance status and reproductive success are both age-related, so that dominance and reproductive success are only incidentally correlated. This interpretation cannot be ruled out; in fact it is supported by the available data. Ages are known for 9 of the 12 clearly subordinate females who reproduced (mean = 3.7 years, SD = 1.4) and for 10 of the 14 clearly dominant females (mean = 5.0 years, SD = 0.4). The age difference between subordinate and dominant females is significant.[10] Thus, older individuals are more likely to be dominant, and dominant individuals have a higher reproductive success than do subordinates. Not surprisingly, a similar age-related pattern of social dominance appears to hold for yellow-bellied marmots as well: the percentages of submissive responses by two- and three-year-old female marmots were

[7]binomial test, P < .05
[8]binomial test, P < .05
[9]Fisher's exact test, P < .05
[10]Mann-Whitney U test, P < .05

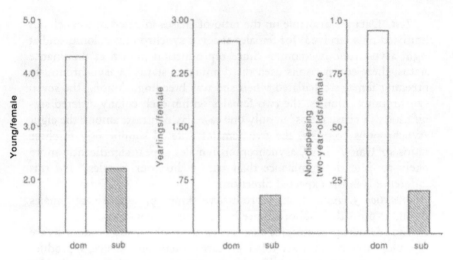

Fig. 9.5. Reproductive success of dominant and subordinate *M. caligata* females. Based on 14 dominant and 12 subordinate females.

49, 57, 58, and 75; for four-year-olds, 30, 34, and 45; and for five-year-olds, 0 (Armitage, 1965). It is also no great surprise that adult female woodchucks consistently dominate yearling females (F. H. Bronson, 1964). I observed five interactions between adult females and yearling females; the adult was clearly dominant in all five cases.

The apparent role of age in mediating dominance status does not in any way contradict the apparent role of female/female competition in mediating reproductive suppression. Although it would be profitable, eventually, to disentangle these influences—perhaps by obtaining a sufficient sample of young, dominant females and old, subordinate ones—until that time we can nonetheless conclude that dominant female hoary marmots have a higher reproductive success than do subordinate females, regardless of the specific proximate mechanisms that lead to this difference.

In summary, hoary marmots reveal a syndrome of reproductive patterns suggesting that suppression of reproduction of and by females is an important aspect of their biology. Reproductive skipping is more pronounced in cases that would otherwise lead to synchronous breeding with the co-resident female; reproductive asynchrony correlates with disparity in social rank; and finally, subordinate females appear to leave fewer descendants than dominants, both because they are occasionally inhibited from reproducing when co-resident females are doing so and because when they do reproduce, they may nonetheless be partially suppressed.

A similar pattern may well hold for other marmot species. As already described, yellow-bellied marmot females are more likely to skip an additional reproductive year if they occupy larger harems (Downhower and Armitage, 1971). In addition, adult females that were subordinate to a clearly dominant female failed to have young emerge, although they apparently became pregnant; by contrast, two females in the same study area which did not interact directly with this same dominant female both produced litters (Armitage, 1965). The following data (from Armitage, 1986b), illustrate the effects of living solitarily as opposed to living near other adult females on weaning success of pregnant female yellow-bellied marmots aged three or older:

	Solitary	Near others
Litter weaned	14	20
No litter weaned	2	20

The difference is significant.[11] Moreover, the reproductive success of two-year-old females is affected by the presence of adults, even relatives. If two-year-old yellow-bellied marmots reproduce at the same rate as adults, 50% should do so. When adult females were present, however, 14 of 48 two-year-olds reproduced.[12] When their mother was present, only 11 of 36 two-year-olds reproduced,[13] whereas when no adult females were present, 10 of 26 two-year-olds reproduced, a value not significantly different from 50%[14] (Armitage, 1986b). Reproductive suppression was most apparent when harems were larger than two and one female was part of a matriline consisting of only herself. In such cases, the single-female matriline averaged only 0.37 litter per year, as opposed to 0.58 litter/year for monogamous females and 0.61 litter/year when the matriline consisted of two females (Armitage, 1986b). A female is apparently inhibited from breeding when she resides with two or more females that are closely related to each other. This inhibition is absent when each female is "equal," i.e., when neither one is associated with additional relatives.

In general, however, it may be that reproductive suppression is not a pronounced phenomenon in yellow-bellied marmots, probably for the same reason that density-dependent effects generally are less evident in that species: because of their reduced social tolerance, individual yellow-bellied marmots encounter each other less often than do Olympic or

[11] chi-square, $P < .01$
[12] chi-square, $P < .01$
[13] chi-square, $P < .001$
[14] chi-square, $P > .10$

hoary marmots. Hence, the opportunity for reproductive suppression is diminished—as is, perhaps, the adaptive significance of such suppression.

Since woodchucks interact less frequently than do yellow-bellied marmots, evidence regarding female/female reproductive suppression is correspondingly even more difficult to obtain. Of course, for the same reason, female/female reproductive suppression may also be less important among woodchucks. I obtained data similar to the adult neighbor sex ratios presented earlier, characterizing adult female woodchucks by the local density and sex ratio of the resident animals surrounding them. I followed the reproductive histories of eight different adult females and six different yearling females and indexed them as to (1) the number of other adults and yearlings resident within a plot of 4 hectares centered on the focal individual, and (2) the ratio of females to males among these neighbors.

The results are as follows: all eight females reproduced, and there was a weak but positive correlation between litter size, which ranged from 1 to 5, and number of neighbors, which ranged from 1 to 7.[15] However, there was no effect of female:male neighbor ratio upon the litter size of focal females.[16] Of the six focal yearling females, three reproduced and three did not. All three of the former produced two young apiece, so the admittedly meager sample yielded no suggestions of correlations between litter size and local competition. However, the three reproducing yearlings had three, four, and five neighbors whereas the three nonreproducing yearlings had one, two, and two neighbors, suggesting that yearlings are *more* likely—not less—to reproduce when they have neighbors. The neighbor sex ratios (female:male) for the reproducing yearlings were 0.5, 0.33, and 0.25, whereas for the nonreproducing yearlings they were 1, 1, and 0.5.

Given the essentially solitary nature of woodchucks, it is not surprising that interindividual reproductive suppression is not prominent. The weak positive correlation between density and reproduction in adults occurs presumably because better-quality habitat tends to attract more residents, which also are more successful reproductively—in all probability, *despite* each other's presence. (It may also be due to the very small sample.) Virtually all adult female woodchucks breed every year; of 174 examined in central Pennsylvania, for example, only seven were not pregnant (Snyder, 1976). Not surprisingly, no reproductive suppression was found, or expected, in the above observations.

[15] Kendall correlation coefficient, $P < .10$
[16] Kendall correlation coefficient, $P > .20$

Only about 20 to 30% of yearling female woodchucks normally breed, however (Davis, Christian, and Bronson, 1964). Because of the enormous fitness advantage accruing to individuals who reproduce a year early (Cole, 1954), we can expect that the evolutionary calculus should carefully weigh the strategic consequences of breeding as a yearling as opposed to deferring reproduction. In a central Pennsylvania population that had been artificially depressed by the removal of potentially reproductive adults, the average percentage of yearling females that were pregnant increased from 20 to 56% (Snyder, 1976). It is tempting to suggest that the "decision" to breed or to defer is influenced by the probability of experiencing competition from other females. If so, then it makes sense that a reduction in such competition would induce more yearlings to come into estrus. More data are needed, however, especially to determine whether the observed compensatory increase in early sexual maturation is due proximately to changes in the endocrine axis as a result of reduced competition and/or ultimately to differential success in scramble competition, irrespective of female/female competition *per se*.

Dominance and Proximate Competition

The immediate, proximate forms of female/female competition influence the behavioral interactions between female marmots. Among the montane species at least, the adult males' strategy typically emphasizes harem acquisition and defense and, among yellow-bellied marmots in particular, territoriality. The adult females' strategy emphasizes benevolence toward sisters and daughters (see Chapter 10) and agonism toward other adult females. Adult female yellow-bellied marmots show more agonistic behavior toward each other than expected from their proportion in a colony population (Armitage and Johns, 1982). Such agonism fluctuates in frequency during the first few weeks following spring emergence, becoming highest around week 4, at the time of parturition (Armitage, 1965). Female/female agonism is highest when home ranges overlap over several weeks following emergence; it diminishes as individual females spread themselves out across the meadow and establish separate home ranges, making interactions between them significantly less frequent.

Developing young yellow-bellied marmots tend to assume their mother's home range. As a result, interaction with other adult females is reduced; in particular, as yearling females concentrate their activities in areas frequented by their mother and not by other adult females, they experience fewer interactions with the latter and so are less exposed to both the proximate and ultimate effects of female/female competition. This sheltering effect leads to "spatial buffering" of the maturing females via the pattern of home-range utilization by adult females and their offspring. It is noteworthy that female yearlings that grow up spatially buffered in this way are less likely to disperse and, hence, are more likely to be recruited into their natal colony (Armitage, 1984; see Chapter 13).

It might be expected that the presence of adult females other than the mother would accordingly reduce the frequency with which yearling female yellow-bellied marmots remain in their colony, ultimately to reproduce. But, in fact, this is not the case (Armitage, 1984), apparently because of the efficiency with which spatial buffering protects yearlings from female/female competition. On the other hand, the arrival of an immigrant adult female correlates strongly with the dispersal of yearling females from their natal colony (Armitage, 1984). Adult females immigrate successfully when a resident adult female (mother of the yearlings) has died or emigrated. As a result, the yearlings are not spatially buffered from female/female competition, and, significantly, dispersal is more likely. Unlike the impact of immigrant adult females, the presence of an immigrant adult male does not influence the dispersal tendencies of female yearling yellow-bellied marmots.

A similar pattern occurs among Olympic and hoary marmots, except that individual females are less aggressive and members of the same colony typically do not isolate themselves as distinctly from one another. Among Olympic marmots, for example, nonreproductive adult females, adult males, two-year-olds, and yearlings may all enter the littering burrow of a reproductive female, even one that is apparently quite dominant. This is less frequent among hoary marmots, but even in this species, overt signs of female/female intolerance are not so clear as to be manifested in discrete home ranges for each adult female, as in M. flaviventris. Among woodchucks, adult females' home ranges may overlap, but interactions among females are even less frequent than among yellow-bellied marmots, although a similar seasonal pattern of agonistic behavior takes place, with agonism highest during the weeks immediately following spring emergence (F. H. Bronson, 1964). It seems likely that during this time, potentially reproductive female marmots of all species compete in varying degrees—for status among the more social species, for territories or home ranges among the more asocial—factors that eventually influence reproductive success.

In very rare cases, female/female competition for reproductive success may be overt, including infanticide. Armitage, Johns, and Andersen (1979) noted three instances of cannibalism during 16 years of research on

M. flaviventris. At least one of these cases was clearly infanticide rather than scavenging, since an adult female was seen to kill a young of the year from another litter. In at least one other case, the victim was related to the perpetrator, although the coefficient of relationship was 1/8 (Brody and Melcher, 1985). It seems unlikely that a need for food was responsible, since forage was abundantly available at the time.

Among woodchucks, unlike the montane species, adult females are quite widely dispersed, both at spring emergence and later during the active season. Given their higher reproductive rates and the early dispersal and maturation of their offspring, however, woodchuck females may actually compete no less vigorously among themselves than do the montane marmots, whose communal life style provides more opportunities for overt competition. Indeed, the physical and behavioral separation of woodchucks may be seen as a result of such competition no less than a means of avoiding it.

Evidence suggests, in fact, that female/female competition in woodchucks differs from that in the montane species largely in being prolonged throughout the active season, possibly becoming even more intense than male/male competition from midseason until hibernation. Thus, an examination of the frequency of severe wounding among woodchucks reveals that whereas more males than females are severely wounded during March and April (9% vs. 1%), the situation is reversed in May and June (5% vs. 7%). Moreover, yearlings show a similar pattern: 3% of yearling males are severely wounded in March and April, as compared to 0% of yearling females, whereas by May and June these figures are 0 and 8%, respectively (F. H. Bronson, 1964).

It is noteworthy that dominance relationships between males and females are less predictable among woodchucks than in the montane forms, in which adult males always dominate adult females. To be sure, male woodchucks often dominate females, but not invariably: of 20 paired laboratory encounters, ten were won by males and ten were won by females (F. H. Bronson, 1964). Significantly, sexual dimorphism in body weight is low in woodchucks; for example, Grizzell (1955) reported that among animals live-trapped in Maryland, the females outweighed the males in ten of 22 cases. Sexual dimorphism appears to be less pronounced among woodchucks than among any other known marmot species (Fig. 9.6).

This trend may be due to an increasing tendency toward polygyny in the sequence *M. monax, flaviventris, caligata.* Even though average harem size in *M. caligata* is smaller than in *M. flaviventris,* the former is less permeable to transients and dispersers, perhaps because of the fitness payoff

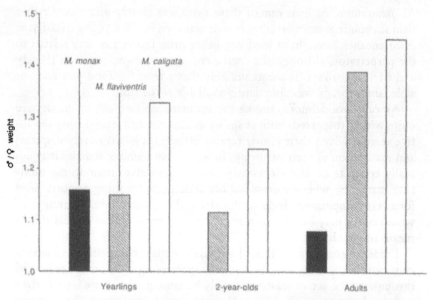

Fig. 9.6. Sexual dimorphism (male/female body weight) for *M. flaviventris*, *M. monax*, and *M. caligata* in June. (Data for *monax* from Snyder, Davis, and Christian, 1961; *flaviventris*, from Armitage, Downhower, and Svendsen, 1976; data for two–year-old *monax* are unfortunately not available.)

to males for being large and successful in male/male competition. On the other hand, the evidence presented above suggests that female/female competition may be quite severe among woodchucks, in which case their low degree of sexual dimorphism may reflect enhancement of female competitiveness rather than a diminution of male competitiveness. Moreover, simple numerical measures of offspring produced are less valid measures of reproductive success than is the production of *successful* offspring, and it seems likely that in most situations nondispersing individuals are likely to be more successful reproductively than individuals that leave their natal colony and almost certainly suffer high mortality (see Chapter 11).

Among yellow-bellied marmots, reproductive status is distributed among resident females somewhat evenly. It may be significant, however, that the production of nondispersing yearling females actually recruited into the breeding population seems concentrated among a relatively few adult females (Armitage, 1984). As a result, it is at least possible that the variance in fitness of adult females is also substantial, possibly comparable to that of the adult males. This possibility must remain unevaluated until students of marmot biology obtain measures of lifetime rather than yearly reproductive success, as well as data on the fitness of dispersing females and males, to complement existing knowledge of the fitness of those nondispersers that are recruited into known colonies.

During my study of woodchucks, I noted that adults tend to be essentially solitary foragers. While feeding, however, individuals occasionally wander close to each other. When they discover each other's presence, the two animals typically separate. Rarely, they remain in proximity. I observed such individuals feeding within 5 meters of each other on 14 occasions: ten of these involved a male and a female, three involved two males (during August), and only one involved two females.

Overlap of females' home ranges is lower in *M. monax* than in *M. flaviventris*, and lower in *M. flaviventris* than in *M. caligata* or *olympus*. Adult females, in turn, appear to be generally less aggressive in this same sequence. Of course, insofar as a high level of aggressiveness results in semisolitary living, this diminishes the opportunity for such aggressiveness.

Summary and Conclusions

1. Female reproductive success varies inversely with harem size, with likely implications for female/female competition.

2. Female aggressiveness increases with pregnancy, again suggesting a competitive component of female reproductive success.

3. Food appears to be the limiting resource that drives female/female competition among the montane species.

4. Hoary marmots reveal a syndrome of reproductive suppression in which females minimize competition and maximize their own fitness by suppressing their own reproduction or that of others in an array of predictable circumstances.

5. A similar pattern of reproductive suppression may well prevail in the other marmot species, but probably with less robustness in *M. flaviventris* than in *M. caligata* and with less yet in *M. monax*.

6. Female/female competition may also be overt, manifesting itself in direct agonistic behavior, with implications for dispersal and recruitment, as well as survivorship of offspring.

7. Female/female competition may also be intense among *M. monax*, but is reflected less in reproductive suppression than in a diminution of sexual dimorphism in both body size and social behavior.

CHAPTER 10

Kin Selection

Inclusive fitness theory (W. D. Hamilton, 1964), more commonly known as kin selection (Maynard Smith, 1964), has emerged as one of the most important recent contributions of evolutionary biology to our understanding of social behavior. Despite some impressive exceptions, however (e.g., Sherman, 1977; Kurland, 1977), there have been surprisingly few empirical demonstrations of its validity among free-living non-human mammals. Nonetheless, as a general principle, there seems little doubt that kin selection "works," in two ways. On the one hand, kin selection provides a precise mathematical statement of the process of natural selection acting on alleles, and as such it appears to be impeccable, although subject to modification and reinterpretation. And on the other, kin selection "works" as a practical matter as well, both for the sociobiologist and for the animals in question: it provides a general, predictive description and interpretation of the widespread tendency of living things to behave preferentially toward relatives.

In particular, inclusive fitness theory offers a coherent explanation for the apparently paradoxical tendency for individuals to behave "altruistically," that is, to act in a manner that enhances the reproductive success of others at some cost to themselves.

According to inclusive fitness theory, alleles for altruism should spread whenever $k > 1/r$, where $r =$ the coefficient of relationship between altruist and beneficiary, and $k =$ the ratio of benefits derived by recipient to cost incurred by altruist, with both costs and benefits measured in units of fitness. As a result, every living thing is thought to behave so as to maximize its "inclusive fitness," the sum of Darwinian fitness (achieved via direct reproductive success) plus the increased reproductive success of genetic relatives that are affected by the behavior of the individual in question, with the importance of each relative devalued proportionately as it is more distantly related. As a result, phenotypic altruism is revealed to be genotypic selfishness.

Variations in any of the above parameters should predictably influence tendencies toward altruism. In practice, however, it is generally difficult to evaluate quantitatively the total costs and benefits of any given act. Moreover, most vertebrates, including marmots, engage in very little behavior that seems clearly altruistic. But because marmots are relatively sedentary, long-lived animals, they certainly experience opportunities to enhance their inclusive fitness via kin-selection effects. And it seems reasonable to extend the concept of selfish vs. altruistic behavior to competitive vs. cooperative behavior, with the resulting prediction that close relatives should fight less and cooperate more than distant relatives or unrelated individuals (Hamilton, 1971).

As we have seen, marmots engage in a range of social behavior, including such basically prosocial acts as mutual grooming and greeting, and agonistic ones such as chasing and fighting. Moreover, patterns of social grouping, the shared use of space, antipredator behavior, and possible reproductive inhibition could provide opportunities for individuals to maximize their inclusive fitness via kin selection.

A Simple Field Manipulation

A major difficulty in studying possible correlations between genetic relatedness and social behavior, in any living thing, is the problem of establishing a high-confidence baseline situation of unrelated individuals. By trapping pregnant female hoary marmots, then releasing them at another colony at which a female has been removed and provisioning that newly introduced female with additional food, I was able to introduce adult females into new colonies. Because of the large distances between these colonies (in some cases, several hundred kilometers) and the highly dissected intervening terrain, there is a vanishingly low probability that the newly introduced females shared a coefficient of relationship with their new colony colleagues that was any higher than that for the species as a whole.

The introduced females reproduced in their new colonies, and their behavior was not distinguishable from that of natural colony residents. True to their biennial breeding schedule, most of these introduced females bred with the resident male two years later. Introductions of this sort were performed successfully at five different bigamous colonies (one in 1978, two in 1979, and two in 1980); in two other colonies, the adult females (known to be siblings) were left intact, but in one of these, the adult male died in 1980. A new adult male—unrelated to either female or to the previous male—was then introduced. He subsequently bred with the resident females. In all, three different categories of offspring inter- action were thereby produced: young/young, yearling/yearling, and young/yearling, and within each of these categories the interactions among individuals of known, differing genetic relationships could be evaluated. Table 10.1 presents the various introductions and subsequent data set thereby produced. The behavioral data presented below were re- corded during 1980, 1981, and 1982, from seven different colonies, and represent 787 total observation-hours. Interactions among full-sib infants and full-sib yearlings all took place among littermates, whereas all other dyads refer to interactions between individuals of different litters.

The introductions generated the following dyads: (1) unrelated young/ young (offspring of introduced female and offspring of resident female, which had different fathers as well); (2) unrelated yearling/yearling (the same individuals, one year later); and (3) unrelated young/yearling (young of the introduced female and yearlings of resident female). During the second year following introduction, young/young paternal half-sib dyads were available (same putative father). And during the third year, young/ yearling and yearling/yearling half sibs had been produced. In addition, the following combinations were available for observation at the two colonies at which no introductions were made and at which the adult fe- males were full sibs: young/young, yearling/yearling, young/yearling half-sib cousins (same father, mothers are sisters), and, following the re- placement of the male at one of these colonies, young/yearling cousin dyads (different fathers, mothers are sisters). Note that no young/year- ling full-sib interactions were observed, since hoary marmot females breed in alternate years only. For most of the observed combinations, se- quential years provided both additional data and a control of sorts: e.g., young/yearling dyads in 1980 were replaced by yearling/young dyads in 1981 as the 1980 young of the year became yearlings and the dam who had yearlings in 1980 produced young of the year in 1981.

The results are summarized and simplified in Table 10.2. Because these data could be biased by the number of individuals in various cate- gories present in a colony at a given time, it is necessary to show that the observed rates of greeting, allogrooming (grooming another individual), fighting, and chasing differed from what would be expected on the basis

TABLE 10.1

Data Set upon Which the M. caligata *Inclusive Fitness Field Experiment Is Based*

Interaction	Genetic relationship	Coefficient of relationship	Number of litters	Number of individuals	Hours observed
Young-young	Unrelated	0	4	15	169
Young-young	Half sibs	1/4	8	27	324
Young-young	Half-sib cousins	3/8	2	8	85
Young-young	Full sibs	1/2	7	22	403
Young-yearling	Unrelated	0	4	11	96
Young-yearling	Cousins	1/8	2	8	81
Young-yearling	Half sibs	1/4	6	18	222
Young-yearling	Half-sib cousins	3/8	2	6	90
Yearling-yearling	Unrelated	0	4	14	153
Yearling-yearling	Half sibs	1/4	4	14	186
Yearling-yearling	Half-sib cousins	3/8	4	15	87
Yearling-yearling	Full sibs	1/2	9	30	325

of chance alone. Accordingly, expected frequencies were calculated after the method of Altmann and Altmann (1977). The results were as follows:

1. Full-sib young/young and yearling/yearling dyads—both of introduced and unmanipulated resident females—engaged in significantly more greetings and allogrooming than any other category,[1] more than would be expected from their frequency in the population.[2] However, they also engaged in significantly more fighting and chasing,[3] and they spent more time in close proximity.[4] Again, these categories exceeded expectations based on chance.[5] There was no difference between the full-sib interactions among the offspring of introduced and resident females.[6]

2. Although some statistical differences distinguished young/young, young/yearling, and yearling/yearling dyads, within each dyad type there were no differences in social behavior between any of the other categories of genetic relatives. The frequencies of interactions among young/young, young/yearling, and yearling/yearling dyads also did not differ from the frequencies predicted on the basis of chance alone, given the abundances of each genetic category in the population.[7] In other

[1] t test, P < .01
[2] chi-square test, P < .01
[3] Kruskal-Wallis rank sum tests for multiple comparisons, P < .01
[4] binomial tests, P < .01
[5] chi-square test, P < .01
[6] chi-square test, P > .10
[7] chi-square test, P > .05

TABLE 10.2

Behavioral Data for Different Dyads

Interaction	Genetic relationship	Greetings		Events per individual/observation-hour			Proximity (%)
		Number	Seconds	Allo-grooming	Fights	Chases	
Young-young	Unrelated	0.23	3.2	0.04	0.08	0.003	11
Young-young	Half sibs	0.14	1.7	0.07		0	9
Young-young	Half-sib cousins	0.17	2.0	0.01	0.04	0.002	16
Young-young	Full sibs	1.79	0.7	0.76	0.43	0.45	66
Young-yearling	Unrelated	0.20	1.0	0.32		0.04	6
Young-yearling	Cousins	0.16	1.1	0.24	0.01	0.02	4
Young-yearling	Half sibs	0.15	0.9	0.31	0	0	5
Young-yearling	Half-sib cousins	0.12	1.4	0.21	0	0	5
Yearling-yearling	Unrelated	0.46	2.7	0.14	0.43	0.36	15
Yearling-yearling	Half sibs	0.39	2.5	0.16	0.52	0.40	18
Yearling-yearling	Half-sib cousins	0.44	2.4	0.16	0.38	0.45	14
Yearling-yearling	Full sibs	1.52	1.0	1.22	1.55	0.97	52

NOTE: Greetings/seconds = duration of observed greetings. Proximity = % of censuses—taken every 10 minutes—at which individuals were less than 4 meters apart. Proximity comparisons were based on binomial tests; all other comparisons based on Mann-Whitney U tests.

words, with the exception of full sibs, other categories did not vary their social interactions as a function of their genetic relationship. Therefore, this limited field experiment did not reveal any apparent tendency among hoary marmots for closer relatives, except for full sibs, to behave more affiliatively and less agonistically.

These results were unexpected. Even now, they clearly do not provide a conclusive failure to confirm inclusive fitness theory among hoary marmots, in that they can be explained (cynics might say "explained away") by any combination of the following possibilities:

1. Perhaps hoary marmots simply lack the ability to discriminate different degrees of genetic relatedness, beyond the full-sib relationship that is coextensive with one's littermates. Kin-biased behavioral discrimination seems to be a reasonable expectation for the genus *Marmota*; since its demonstration would have been considered supportive of inclusive fitness theory, its apparent absence requires an explanation.

2. Perhaps kin discrimination is entirely based on familiarity; thus, marmots generally live with close relatives: sibs, nieces, nephews, rarely relatives more distant than cousins. Accordingly, social familiarity may serve as a reliable and useful cue to genetic relatedness and as a basis for kin recognition.

3. Perhaps hoary marmots did not discriminate as expected because they normally do not encounter individuals of sufficient genetic variability to have been selected for such discrimination. Indeed, it appears that both *M. olympus* and *caligata* experience less gene flow than does *M. flaviventris* (see Chapter 13). And even among yellow-bellied marmots, individuals treat others related by $r < .5$ as nonrelatives (Armitage and Johns, 1982), even though the opportunity certainly exists for kin preferences to be expressed toward other, more distant relatives. If, during their evolutionary history, hoary marmots typically have had little opportunity to interact with nonrelatives or distantly related individuals, then selection may not have favored such discrimination.

Another well-studied sciurid rodent, the Belding's ground squirrel (*Spermophilus beldingi*), does not discriminate individuals more distantly related than $r = .25$, apparently because members of this species lack both demographic and geographic opportunities to enhance their inclusive fitness by interacting with more distant relatives (Sherman, 1980). Hoary marmots, however, are relatively long-lived, commonly surviving to seven years of age and older, so demographically, at least, opportunities for fitness-enhancing interactions with distant relatives seem likely to exist; even with a high overwinter mortality of .60, for example, the opportunity for paternal half sibs to encounter each other is .16. It remains to be seen whether restricted gene flow in this species is responsible for the observed lack of kin-oriented discrimination.

Comparing *M. monax*, *flaviventris*, *olympus*, and *caligata*, we might expect to find a cline in the amount and diversity of gene flow normally encountered, with woodchucks probably experiencing the most. They are most likely to encounter dispersing individuals, some of which are quite distantly related, if at all. Because they occupy more discontinuous habitats and constitute a genetically more viscous population than does *M. monax*, yellow-bellied marmots should encounter a reduced genetic variance in their social environment (Schwartz and Armitage, 1980). And Olympic and hoary marmots, with their slower reproductive rates and relatively high social isolation and low frequency of dispersal, should encounter the least genetic diversity. A further prediction, accordingly, is that these species should reveal a parallel cline in social tolerance, as in fact they do.

This is not to suggest that the observed interspecific gradient in social tolerance is due solely to differences in population viscosity, or even that the latter is necessarily a significant factor contributing to the former; rather, the two patterns are consistent with one another.

Paradoxically, reduced gene flow among hoary marmots could be partly responsible for both the observed higher level of social tolerance via kin selection and, at the same time, the observed failure of individuals to achieve precise discrimination among relatives, if hoary marmots typically do not experience sufficient genetic diversity among prospective interactants to have been selected for such discrimination. This argument, although plausible, is nonetheless troublesome, since consistent exposure to even a limited range of genetic relatives—as is apparently the case for hoary marmots—should select for appropriate discrimination within that typical range. Resolution of this question will require more precise information on rates of gene flow within each species.

4. It is also possible that my failure to find results consistent with kin-selection theory in this case resulted because the behaviors here reported simply are too inconsequential for selection to have operated on them. Perhaps the categories "grooming," "greeting," "fighting," and "chasing" do not warrant the implications for inclusive fitness herein attributed to them; i.e., the costs and benefits of such acts may be quite small.

5. Finally, perhaps the conclusions represent a Type II error—accepting the null hypothesis when it is actually invalid. Thus, additional data may conceivably demonstrate a trend that is not apparent from the results presented here.

Inclusive fitness theory is clearly too robust, in both theoretical and empirical support, to be seriously undermined by any single set of findings, especially those involving so small a sample as the above. Subsequent work is clearly more likely to modify our understanding of marmot social dynamics than to modify inclusive fitness theory. And as we shall see in the following sections, inclusive fitness theory sheds much useful light on marmot social dynamics. Nonetheless, this field experiment was conducted in the full expectation of confirming predictions based on this theory; had the predicted results been obtained, I would have enthusiastically reported them as a confirmation. Accordingly, they must now be acknowledged as a failure to confirm.

Alarm Calling

Marmots are highly vocal, with the exception of the woodchuck, which, significantly, is both the most silent and the least social. The most notable marmot vocalization, the long alarm call, lends itself to interpretation within the framework of inclusive fitness theory. Thus, by responding to a potential predator by giving an alarm call, the caller informs other marmots of an existing danger, thereby (presumably) increasing the chances that others will survive and ultimately reproduce, while at the same time (presumably) decreasing its own chances of survival and its own ultimate evolutionary success. Accordingly, such behavior appears to qualify as "altruistic." Kin-selection theory therefore suggests that it should be preferentially oriented toward relatives, although other possible evolutionary factors must also be considered (e.g., reciprocity: Trivers, 1971).

Imagine an allele that induces its carrier to give alarm calls when a predator is sighted, as opposed to its alternative allele, which induces its carrier to remain silent. If alarm calling does in fact reduce the caller's immediate success while increasing that of other individuals, then alarm calling is an important potential paradox for evolutionary theory; that is, a behavior which is truly altruistic should result in a *decrease* in frequency of the corresponding encoding alleles, and, therefore, it should disappear. Its persistence among marmots suggests, although it does not prove, that (1) the behavior in question does not actually reduce the survival of the alarm caller (i.e., it might be selfish or neutral, in a variety of ways), or (2) it does not increase the survival of those hearing the call (once again, it might be selfish or neutral), or (3) the benefit of such behavior may reside in advantages it confers on the entire colony of marmots, rather than on individuals (i.e., group selection as espoused originally by Wynne Edwards, 1962, and more recently by a variety of more mathematically inclined workers), or (4) the benefit of alarm calling may reside in the indirect or kin-selected benefits accrued by the alarm caller via its relatives who profit from the alarm calling. It is also likely that the situation is complex, not simply attributable to any of these factors acting alone.

By calling, marmots doubtless increase their own conspicuousness. Often, I have become aware of an active colony by hearing the alarm calling generated by my presence. Although individual marmots clearly draw attention to themselves when they call, it is less clear that alarm calling actually increases the predation risk to which they are subjected. Thus, marmot alarm calls have characteristics like those of avian alarm calls, which have a ventriloquial effect and make the caller difficult to locate (Marler, 1957): high frequency, with abrupt beginning and end, and little frequency modulation. These apparent antipredator adaptations of marmot alarm calling can be interpreted either as confirmation of the dangerousness of calling—and hence of its likely altruism—or of its safety!

Observations suggest that whereas some risk is associated with marmot alarm calling, that risk is not very high. Accordingly, if alarm calling is altruistic, it is not "very altruistic." In 18 years' observations of marmots, I witnessed 13 cases of successful predation; in every case, the victim was taken apparently unaware, before it or anyone else gave alarm calls. Field experiments using a radio-controlled model golden eagle elicited numerous alarm calls from a free-living population of hoary marmots in the Washington Cascades (Noyes and Holmes, 1979): when marmots were more than 5 meters from cover at the time the "predator" was spotted, all individuals (17) called and simultaneously ran to cover; all marmots (6) that were less than 1 meter from cover called and visually tracked the intruder; all individuals (6) that were observed to be in an exposed meadow when they spotted the "predator" gave alarm calls as they ran. These latter observations suggest either that alarm calling does not significantly increase the risk to the caller (so that individuals even call when they are exposed) or, alternatively, that the benefit of calling is so great that alarm callers will do so even before attaining a safe refuge.

Among European Alpine marmots, adult males do significantly more alarm calling than do the members of any other age/sex class (Barash, 1976a; Table 10.3).[8]

In this species, males characteristically remain 100–200 meters above

[8] t test, P < .05

TABLE 10.3

Frequency of Alarm Calling Sequences per Animal
per Observation-hour for M. marmota

	June		July		August
	1–15	16–30	1–15	16–31	1–15
Adult males	1.80	0.96	1.52	1.60	2.40
Adult females	0.44	0.51	0.20	0.33	0.65
Yearlings	0.29	0.04	0.16	0.16	0.23
Young	—	—	0	0	0.01

NOTE: Observed in Vanoise National Park, France, 1973. Based on a minimum of 55 hours in each 2-week interval.

the other colony members. This positioning may be due to a degree of antipredator vigilance; it also facilitates their detecting other males seeking entrance to their colonies (Chapter 8), so their higher alarm-calling rate may be an accidental consequence of their vigilance against other gallivanting males rather than kin-selected altruism. In this respect, however, it is notable that the higher rate of alarm calling by adult male Alpine marmots continues into August, at which time anti-gallivanting behavior should have diminished.

I knew the identity of the initial alarm caller for 117 cases of alarm calling by hoary marmots in response to a potential predator. Adults and two-year-olds combined were responsible for 71; yearlings were responsible for 26; and young of the year were responsible for 20. Comparing these data on the basis of alarm calls per individual present per observation-hour reveals a significant trend: older animals do more alarm calling than younger ones,[9] and they do more alarm calling than would be expected from their proportion in the population.[10] Such findings are consistent with the expectations of kin-selection theory, since older animals are more likely to have kin (both direct descendants and collateral relatives) in the population and hence are more likely to increase their inclusive fitness by raising the probability that other colony residents will survive to reproduce. On the other hand, the higher frequency of alarm calling by older animals may be an incidental by-product of greater alertness because of heightened intraspecific social salience with increasing age, in which case it is most parsimoniously interpreted as "selfish." Older animals are also larger; hence they may be less susceptible to predators. If so, their higher frequency of alarm calling may be less costly for

[9]Kruskal-Wallis analysis of variance, P < .01
[10]chi-square test, P < .05

them than for younger individuals, and, accordingly, alarm calling by them may be less altruistic.

A four-year study of hoary marmots (Barash, 1980a) revealed that reproductive females averaged 0.80 (SD = 0.3) alarm calls per individual per hour, whereas nonreproductive females averaged 0.33 (SD = 0.2). The difference is significant,[11] and may well reflect greater parental watchfulness. Additional differences in behavior between reproductive and nonreproductive females are interpretable as reflecting parental investment by the females in question (see Chapter 12) and should be distinguished from kin selection, the latter being most usefully restricted to fitness maximization via influences on relatives other than dependent offspring.

If alarm calling reflects kin selection—as well as parental effort—then newly immigrant adult males, with no relatives in their colonies, should do less alarm calling than returning males, who have sired offspring. There is a tendency in the predicted direction, but the difference is not significant: returning male hoary marmots averaged 0.79 calls/individual/hour (SD = 0.32) whereas immigrant males averaged 0.70 (SD = 0.33).[12] However, colony males—more likely to have relatives within their colonies—do significantly more alarm calling than satellite males (0.77, SD = 0.31 vs. 0.35, SD = 0.22).[13] It is also possible, of course, that colony males are more watchful than satellite males for other reasons or that their colony locations afford better vantage and, hence, a higher predator detection frequency.

In summary, data on marmot alarm calling are generally consistent with predictions based on inclusive fitness theory, but they are neither without exception nor strongly persuasive. And given the number of variables that must be considered as well as the low incidence of presumably crucial interactions with predators, field experiments with model predators or trained live predators seem crucial (e.g., Hoogland, 1983).

The Social Use of Space

As we have seen, both Olympic and hoary marmots often associate in foraging groups and walk side by side through the meadows, eating close to each other. Such associations appear to represent "selfish herd" phenomena (W. D. Hamilton, 1971) and/or the enhancement of antipredator

[11] t test, P < .05
[12] t test, P > .10
[13] t test, P < .01.

vigilance by social grouping, as well as (possibly) the simple aggregative effect of clumped food resources. As such, however, foraging groups need not be composed of relatives. Indeed, if such herds are truly selfish, then, from the perspective of each participant, unrelated individuals might even be preferable constituents. Antipredator vigilance could easily be as high among nonrelatives as among groups of genetic relatives, since we assume that each individual is vigilant primarily for its personal benefit, with any benefit derived by other group members presumably an incidental side effect of the selfish behavior of each individual. However, foraging groups—at least of *M. caligata*—are significantly skewed toward genetic relatives.

I randomly sampled 50 foraging groups, ranging in number from four to eight individuals, and defined a foraging group as one in which no member was more than 4 meters away from another member. The average coefficient of genetic relationship among such group members was .38 (SD = 0.10); by contrast, the average coefficient of genetic relationship among all individuals present in the same colony at the same time was .21 (SD = 0.04).[14] On the level of ultimate causation, individuals within a foraging group composed of close relatives might be somewhat better protected if vigilance and alarm calling are greater when relatives are nearby. On a proximate level, foraging groups may tend to be composed of genetic relatives simply because such groups are the likely above-ground extension of patterns of burrow sharing and individual familiarity (which as we have seen are strongly influenced by genetic relatedness).

Food sharing might also be more likely among related members of a foraging group. I experimentally distributed 85-gram servings of peanut butter on 20 separate occasions. Ten times this food attracted more than one individual, and of these, four resulted in brief instances of growling and chasing; in the other six, two or more individuals ate together without apparent agonism. In the former cases, the mean coefficient of genetic relationship among the competitors was .18; for the six instances of amicable food sharing, the genetic relationship averaged .42. It should be pointed out, however, that marmot food sources in nature are considerably more dispersed and, hence, these data may simply be an artifact, possibly reflecting a general correlation between genetic relationship and amicability but not necessarily reflecting an adaptation to food sharing *per se*.

Among yellow-bellied marmots, full sibs and parents/offspring are

[14] Kruskal-Wallis analysis of variance, P < .01

more amicable and less agonistic than are distant relatives (Armitage and Johns, 1982). Kinship patterns are important in determining how much foraging area individuals share, with spatial overlap being greatest among close kin: mothers and their young, and littermates (both as young of the year and as resident yearlings) had virtually identical foraging areas (Frase and Armitage, 1984). Burrowmate yellow-bellied marmots are also more amicable and less agonistic than are individuals inhabiting different burrow systems, and this too is reasonably interpreted as reflecting the influence of kin selection (Armitage and Johns, 1982).

Among Olympic and hoary marmots, the most consistent burrowmates are mothers/young, young/young sibs, and yearling/yearling sibs. As a result, the average coefficient of relationship among burrowmates in July, for the 82 cases for which such data are available, is .41 (SD = 0.04). By contrast, comparable data for 82 randomly chosen residents of different burrows yielded a figure of .29 (SD = 0.19).[15] As pointed out earlier, however, burrow associations are not immutable and are typically more predictable and more consistent with genetic relationship earlier in the season. Thus, for *M. caligata*, the mean *r* for burrowmates in June was .39 (SD = 0.04), by August, .28 (SD = 0.12).[16] Moreover, the standard deviation of *r* among burrowmates was also significantly higher in August than in June.[17] This decline in kin-oriented residence patterns is consistent with the seasonal shift in marmot behavior, from reproduction and other social activities early in the season to more self-involved feeding during the period of intense fat deposition prior to hibernation.

Comparable data for woodchucks show an even stronger tendency for burrow sharing to correlate with genetic relationship. Thus, I never found unrelated individuals sharing a burrow after May 1, by which time breeding was complete. In 79 such cases, burrow residents were solitary; in 27 cases, more than one individual was found to occupy a burrow. In all of the latter instances, the burrow-sharers were either a mother and her young of the year (*n* = 23) or full-sib young in the initial stage of dispersal (*n* = 4). As a result, my observations of *M. monax* show a coefficient of genetic relationship among burrowmates of .5 (SD = 0), which of course is higher than that observed for *M. caligata* or *M. flaviventris*. This is consistent with expectation: within this solitary and highly aggressive species, burrow "sharing" is virtually nonexistent, limited to the brief male/female pairing and the mother/young dyad. Only indi-

[15] t test, P < .01
[16] t test, P < .10
[17] F-max test, P < .01

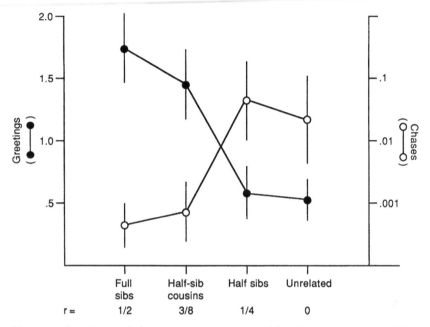

Fig. 10.1. Greetings and chases among two-year-old female hoary marmots (*M. caligata*), as a function of their genetic relationship. Data are presented as number of behaviors per animal per hour, and are based on a minimum of 83 observation-hours per dyad and a minimum of six different individuals in each case.

viduals that are very closely related, it seems, will commonly share the same burrow, and then only for a very brief time. By contrast, among the other, more socially tolerant species, burrow sharing is much more frequent and, when it occurs, is not limited to mothers and young, and full sibs.

The introduction experiment on *M. caligata* described above failed to reveal predicted correlations between patterns of social interactions and gradations in coefficients of genetic relationship among young/young, young/yearling, and yearling/yearling dyads. However, the expected correlation did occur among nonmanipulated two-year-old female hoary marmots (Fig. 10.1). There was a significant positive correlation between genetic relationship and amicable behavior and a significant negative correlation between genetic relationship and agonistic behavior among full sibs ($r = .5$), half-sib cousins (same father, mothers are full sisters, $r = .375$), maternal half-sibs (different fathers, mothers are full sisters,

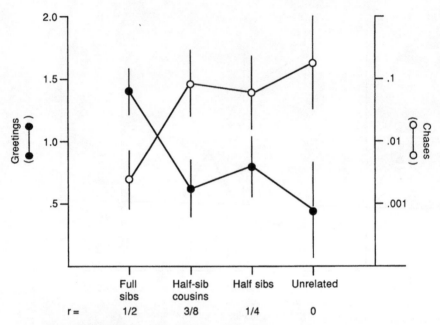

Fig. 10.2. Greetings and chases among adult female hoary marmots (*M. caligata*), as a function of their genetic relationship. Data are presented as in Fig. 10.1, and are based on a minimum of 101 observation-hours per dyad and a minimum of five different individuals in each case.

$r = .25$), and unrelated two-year-olds ($r = 0$). The correlations are significant for both behavioral measures.[18]

A similar pattern is revealed by adult females as well, i.e., individuals known to be three years old or older (Fig. 10.2), although the correlations are less precise. Thus, for adult female hoary marmots there are no trends among individuals less closely related than full sibs.

By contrast, two-year-old male hoary marmots showed no correlations (Fig. 10.3). Because of the social structure of hoary marmot colonies, no comparable data are attainable for adult males. Alternatively, this structure can be seen as indicating a pattern in itself: adult males are not amicable and are vigorously agonistic, the male/male competition for personal reproductive success overriding any possible modulating effects of genetic relatedness.

The correlations between social behavior and genetic relatedness among female two-year-olds and adults are themselves closely associated

[18] Spearman rank correlation coefficients, $P < .05$

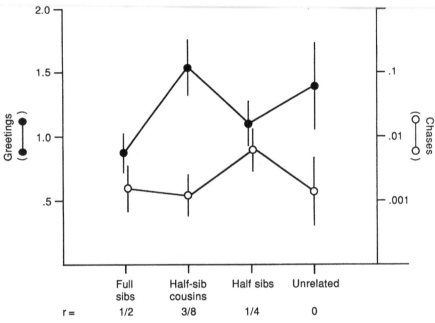

Fig. 10.3. Greetings and chases among two-year-old male hoary marmots (*M. caligata*), as a function of their genetic relationship. Data are presented as in Figs. 10.1 and 10.2, and are based on a minimum of 78 observation-hours per dyad and a minimum of five different individuals in each case.

with patterns of burrow residence, and may proximately be due to such patterns and/or socialization during development. Thus, full-sib two-year-olds and adults are significantly more likely to share burrows than are half-sib cousins or half sibs;[19] the latter, in turn, show a (statistically nonsignificant) tendency to share burrows more than do unrelated individuals of the same age (Fig. 10.4).[20] There is no difference between half-sib cousins and half sibs in this respect, for either adult or two-year-old females.

The above findings suggest that correlations between genetic relatedness and social behavior may actually be due to correlations between genetic relatedness and burrow occupancy. Accordingly, I calculated partial correlations and found that the initial correlation between genetic relatedness and social behavior was no longer significant after the effect of burrow occupancy was removed. I also used multiple regression to examine

[19] binomial tests, $P < .05$
[20] binomial tests, $P < .10$

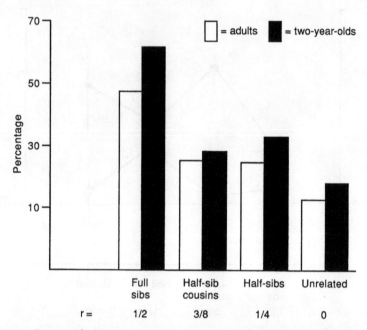

Fig. 10.4. Burrow sharing among adult and two-year-old female hoary marmots (*M. caligata*), as a function of their genetic relationship. Data are presented as a percentage of confirmed burrow occupancy at which co-residence was also confirmed and are based on a minimum of 10 different individuals in each case and 12 different confirmations of nightly burrow occupancy for each individual.

the individual and combined effects of genetic relatedness and burrow occupancy on amicable and agonistic behaviors. The previously significant relationships of each factor disappeared, because genetic relationship and burrow occupancy were strongly intercorrelated. The two factors combined, however, were significantly related to both amicable and agonistic behaviors, accounting for 79% of the observed variance. Unfortunately, sufficient data on burrow occupancy of two-year-old males were not available for comparable analysis, although the trend appears to be similar.

These findings are suggestive but certainly not conclusive. Although chasing is clearly agonistic and is likely to reflect competitiveness—the opposite of altruism—the interpretation of greeting behavior is more problematic. Greeting seems to be primarily amicable and affiliative, although agonistic components are also present, especially among older animals. Measures of greeting, therefore, are not readily translatable into measures of altruism. It should not be surprising that burrow occupancy correlates with genetic relatedness, nor that burrow occupancy also correlates with social behavior so that with the effect of burrow occupancy removed, no significant correlation between genetic relatedness and social behavior remains. It suggests that burrow occupancy and hence social familiarity (Bekoff, 1981; Holmes and Sherman, 1982) are the proximate mechanism through which kin-directed behavior is mediated and the maximization of inclusive fitness is ultimately achieved.

Not surprisingly, burrow sharing correlates with proximity and frequency of social interactions, and, therefore, with social familiarity. Among hoary marmots, for example, burrow residents greeted one another significantly more often than did nonburrow residents, even when the data were corrected for relative availability of individuals of each type.[21] Furthermore, burrowmates spent a higher proportion of time within 5 meters of each other than did nonburrowmates,[22] largely because of time spent lounging about at the burrow entrance at either the beginning or the end of activity periods. Interactions *within* the burrows, of course, provide additional—and unobserved—opportunities for interactions among burrowmates. This, in turn, would seem to provide an open niche for some pioneering and technologically minded aspiring marmoteer: devise and employ an underground burrow-snooper, perhaps a long, flexible fiber-optic tube.

Interactions between residents of different colonies might be expected to differ substantially from those within a colony, because, in the former case, the animals are less familiar with each other; nearly always, they are also less closely related. (Although reciprocity has not yet been investigated among any marmot species, it should also be pointed out that colony residents have greater opportunities for reciprocation as well.) Both Olympic and hoary marmot colonies show considerable variation in colony spacing, and a similar pattern appears to hold for yellow-bellied marmots as well. Whereas some colonies are quite isolated, such that interactions between different colonies are rare, others occupy essentially continuous habitat, with colony borders being virtually contiguous. Even in such cases, colony membership is readily apparent for all individuals except for satellite male Olympic and hoary marmots, which often distribute themselves in the interstices between colony ranges.

Gallivanting colony males interact with individuals from adjacent colonies much less often than they meet their fellow colony members: among Alaskan hoary marmots, for example, about 95% of all social

[21] t test, P < .01
[22] binomial test, P < .01

interactions occurred within each colony, although six colonies were essentially contiguous and not separated by physical barriers (Holmes, 1984b). It is also of special interest that the biasing of interactions in favor of colony members is reversed when it comes to agonistic behavior: at a hoary marmot "colony town" in Washington State, more than 65% of all chases involved residents of different, adjoining colonies, even though interactions between such individuals constituted less than 10% of all dyadic interactions that I observed.[23]

Occasionally, residents of distant, nonadjacent colonies encounter one another. Such animals are clearly less familiar to one another than are colony residents or the residents of adjacent colonies: greetings between these relative strangers are distinctly more vigorous and prolonged, and upright fights and chases are more frequent following such encounters. Thus, 12 such greetings between strange Olympic marmots resulted in nine upright fights (75%) and six chases (50%), while 68 greetings between residents of neighboring colonies resulted in only 20 upright fights (34%) and nine chases (13%). The difference is significant.[24] It may be that proximity discourages overt aggression by facilitating the establishment of more distinct dominance/subordinance relations or that familiarity itself leads to increased tolerance. Whatever the proximate mechanism, aggression between unrelated or distantly related individuals generally exceeds that between closer relatives, and living arrangements—whether burrow sharing or colony residence—seems crucial.

The low level of social integration among woodchucks correlates, not surprisingly, with their scattered distribution and very low frequency of burrow sharing. Cause and effect is difficult to assign here, and is perhaps meaningless in any case: if woodchucks were more socially tolerant, they presumably would also be more likely to share burrows, and/or if they did so, this in itself might well make them more socially tolerant.

It seems likely that marmots lack an innate ability to recognize genetic relatives, by either phenotypic matching (Holmes and Sherman, 1982) or inborn recognition capabilities. Unlike Belding's ground squirrels, which experience intralitter genetic mosaicism—different males may father different individuals in the same litter (Hanken and Sherman, 1981)—marmot litters are almost certainly fathered by the same male. Hence, marmots experience a selective environment in which the benefit of genetically mediated kin recognition is relatively low, and, accord-

[23] binomial test, P < .01
[24] binomial test, P < .01

ingly, it is not surprising that they apparently employ burrow occupancy as a primary cue for social patterning.

To understand why different marmot species experience different levels of social tolerance and, similarly, to evaluate the adaptive significance of sociality itself, we must look ultimately at the different selective forces operating on individuals of each species. It should also be emphasized that although much social behavior among marmots is kin-biased, it is not necessarily altruistic. Whereas phenotypic altruism (and genetic selfishness) may well be involved in alarm calling, kin-biased social behavior, although nepotistic, may also be beneficial to both individuals involved. When closely related individuals forage together and hence reduce the antipredator vigilance of each, or when they share burrows and hence behave less agonistically toward each other, expending less time and energy on stressful social encounters, it seems likely that each participant, as an individual, profits. I therefore suggest that whereas kin-biased, nepotistic behavior characterizes the social interactions of marmots, it may be expressed through mutualism, not necessarily through altruism.

If so, then the question arises: Why don't unrelated individuals behave similarly, thereby achieving comparable benefits? Perhaps the additional, inclusive fitness benefit that results from enhancing the reproductive success of a relative provides a necessary additional adaptive value to such cooperation, and/or perhaps kin are more likely to cooperate because they have an opportunity to do so; i.e., by virtue of common ancestry, kinship provides a possible "uncorrelated asymmetry" (in the terminology of evolutionarily stable strategies) around which socially biased behavior can cohere.

Population Aspects of Kin Selection

Among yellow-bellied marmots, a male's harem typically consists of one or more matrilines—female kin groups each consisting of one or more adult females and their offspring, including young of the year as well as any nondispersing yearlings (Armitage and Johns, 1982). Behavior within such female/offspring units tends to be relatively benevolent, whereas between such groups, competition is the norm. Interactions between mother and young of the year tend to be more amicable and less agonistic than those between a mother and her yearlings. The most persistent and generally least agonistic relationships occur between mothers and their daughters or between young sisters. Yellow-bellied marmot matrilines often bifurcate in patterns of spatial use and general amicability as the sisters become reproductive (Armitage, 1984).

We might predict that yearling female yellow-bellied marmots should be less likely to disperse when the other adults (notably the adult females) in their colony are closely related to them. This is not the case (Armitage, 1984), however, apparently because matrilines effectively share space, thereby protecting younger individuals from competition with distant and unrelated females.

I have already discussed the reduction in reproductive output with increasing harem size among yellow-bellied marmots, interpreting such findings via female/female competition. The observed correlation of large harem size with reduced reproduction per female is also consistent with inclusive fitness theory. Thus, a "large harem" (from the adult male's perspective) consists of two or more matrilineal groups (from the female's perspective). Members of such matrilineal groups, by definition, tend to be more closely related than are members of different groups. Large matrilineal groups suppress the reproduction of small matrilineal groups (Armitage, 1986); in many cases, such "groups" are as small as one female, whose reproduction, as we have seen, is inhibited if that female is outnumbered by other matrilines. It is also noteworthy that whereas female reproductive success is not significantly influenced by an increase in the size of her matriline (that is, by the number of close female-descended relatives with which she is associated), it does decline with an increase in harem size—that is, with an increase in the number of distant relatives or nonrelatives (Armitage, 1986b). Female/female competition and kin selection combine in such cases to exert a potent effect on population recruitment.

For most vertebrates, the fitness payoff is greater for investing in offspring than in collateral relatives (e.g., Rubenstein and Wrangham, 1980); marmots tend to reflect kin-selection patterns in their behavior toward offspring more than toward siblings. Moreover, marmot altruism is typically reserved for younger animals, and behavior among older animals, even if they are related, becomes increasingly agonistic with increasing age, peaking at adulthood.

A similar pattern is found among hoary and Olympic marmots as well. Within the same category of genetic relatives, patterns of amicability decline with increasing age of the individuals concerned. Thus, greeting frequency declines and chasing frequency increases among full-sib young of the year, yearlings, two-year-olds, and adults, with a similar pattern emerging for both males and females (Fig. 10.5).[25] Physical spac-

[25] Spearman rank correlation coefficients, $P < .05$

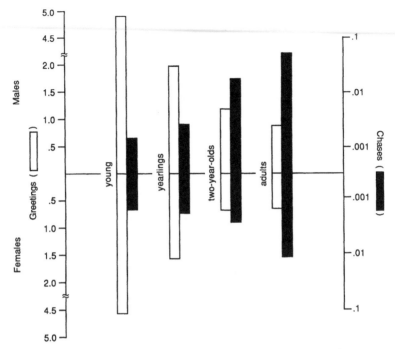

Fig. 10.5. Greetings and chases among full-sib hoary marmots (*M. caligata*). Data are presented per individual per observation-hour and are based on a minimum of eight different individuals in each case and a minimum of 88 observations.

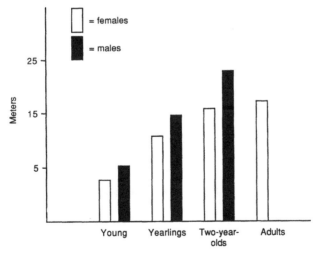

Fig. 10.6. Physical spacing among full-sib hoary marmots (*M. caligata*). Data presented are average distances between individuals, recorded arbitrarily every 5

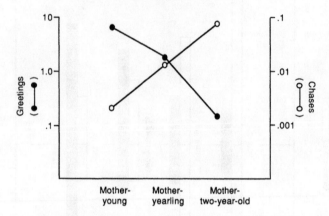

Fig. 10.7. Greetings and chases between mother hoary marmots (*M. caligata*) and their offspring, as a function of age of offspring. Data are presented per individual per observation-hour, and are based on a minimum of 11 different mothers and a minimum of 227 observation-hours for each cohort.

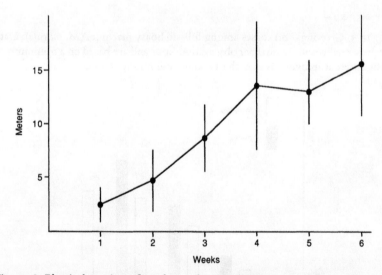

Fig. 10.8. Physical spacing of mother and young hoary marmots (*M. caligata*), as a function of the number of weeks following emergence of the young. Data presented are based on 11 different mothers, with average distances recorded every 10 minutes and a minimum of 122 observation-hours during any one week.

ing between these same individuals also becomes progressively greater with increasing age (Fig. 10.6).[26] A similar trend characterizes mother/offspring interactions, when mothers are compared with their young of the year, yearlings, and two-year-olds (Fig. 10.7). Mothers and two-year-olds greet less and chase more than mothers and yearlings, which in turn greet less and chase more than mothers and young of the year.[27] Once again, physical spacing between mothers and their offspring increases as the offspring age (Fig. 10.8).[28]

On one level, it would seem advantageous for marmot mothers to be maximally tolerant of their young, facilitating close interpersonal spacing and directing little or no aggression toward them. However, individuals that are not genetically identical can be expected to have a differing evolutionary agenda, leading to patterns of competition and conflict (e.g., Trivers, 1974).

Reproductive strategies apparently limit the expression of kin selection, among both yellow-bellied (Armitage and Johns, 1982) and hoary marmots. Thus, among both species, adult males generally behave amicably toward females but agonistically toward other males, including even their sons. As expected, these tendencies are age-related: the reproductive success of adult males is not threatened by young males. As the new males of each developing generation approach reproductive maturity, however, they are increasingly likely to threaten the reproductive success of the existing adult male. Hoary marmots, which do not breed until they are three years old, remain tolerant of their male offspring longer than do yellow-bellied marmots, which become sexually mature as two-year-olds. Whereas aggression from adult males toward yearlings is a prominent feature of yellow-bellied marmot society, as is avoidance of adult males by yearlings (Armitage and Johns, 1982), such aggression and avoidance is rare among Olympic and hoary marmots, but more frequent in interactions between adult males and two-year-olds, especially male two-year-olds (Fig. 10.9).

Adult females tend to behave more amicably toward their daughters than toward their sons. And once again, the trends among hoary marmots resemble those among yellow-bellied marmots, though developing more slowly among the former than among the latter (Fig. 10.10). For young of the year and yearlings of both sexes, mothers behave more amicably and less agonistically than would be expected based on their abun-

[26] Kruskal-Wallis analysis of variance, $P < .05$
[27] Spearman rank correlation coefficient, $P < .05$
[28] Spearman rank correlation coefficient, $P < .05$

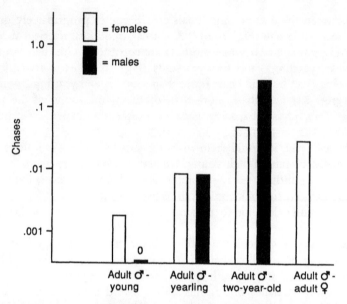

Fig. 10.9. Chase frequencies between adult male hoary marmots (*M. caligata*) and other individuals. Data are presented per individual per observation-hour, and are based on a minimum of eight different adult males in each category and a minimum of 210 observation-hours in each category.

dance in each colony and chance alone.[29] The treatment of two-year-old females does not differ from that predicted by chance,[30] whereas two-year-old males receive significantly more agonistic behavior and less amicable behavior.[31]

Among adult female hoary marmots, interactions parallel genetic relatedness: aunts and nieces interact more agonistically and less amicably than would be predicted by their local abundances, whereas mother/daughter and sister/sister interactions tend to be less agonistic and more amicable than would be predicted.[32] Grandmother/granddaughter interactions occur with a frequency not distinguishable from chance (Fig. 10.11). The situation among yellow-bellied marmots appears to be almost identical (Armitage, 1986a).

Although precise data are not available, it seems likely that adult woodchucks are generally intolerant, regardless of genetic relationship, and that

[29] chi-square test, $P < .01$ for all four cases
[30] chi-square test, $P > .10$ for both amicable and agonistic behavior
[31] chi-square test, $P < .01$ in both cases
[32] chi-square test, $P < .01$ for each

with regard to the ontogeny of parental intolerance, they fit into their "proper" position, beyond *M. flaviventris* and at the other extreme from *M. caligata* (and presumably *M. olympus* as well). Thus, mother/young intolerance develops early and is correlated with dispersal by woodchucks as young of the year (Barash, 1974b) and breeding by yearlings.

The basic patterns of adult sociality among all marmots seem dictated by similar considerations, differing largely in the generations toward which competition is expressed: males eventually seek to exclude other males, whether closely related to themselves or not, while females seek to exclude other individuals that are not closely related. Adult intol-

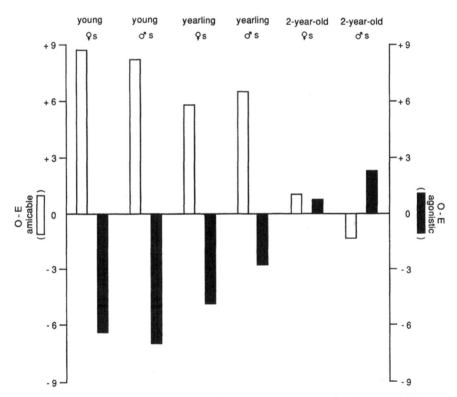

Fig. 10.10. Behavior of adult female hoary marmots (*M. caligata*) toward their offspring. Data are presented as Observed − Expected, with amicable = greeting + grooming, and agonistic = chases + fights. Expected values are calculated after Altmann and Altmann (1977). Based on the observation regime for Fig. 10.7.

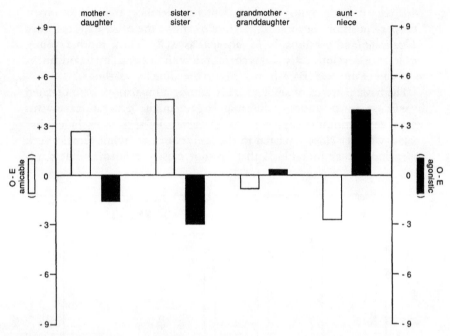

Fig. 10.11. Behavior among adult female hoary marmots (*M. caligata*), as a function of genetic relationship. Data are based on a minimum of 14 different dyads in each category and a minimum of 113 observation-hours in each category.

erance, even toward close relatives, is expressed toward different generations in the various marmot species. Parents behave benevolently toward their offspring, but typically less so as the offspring get progressively older. This trend varies predictably with variation in the age of sexual maturation and, therefore, in the age at which even close relatives become potential reproductive competitors.

The presumed operation of kin selection therefore appears to be limited by the individual reproductive strategies of adults as well as by individual differences among both adults and their offspring (see also Armitage and Johns, 1982). Adult tolerance of offspring is limited by the adults' need to look after their own immediate reproductive interests. As offspring become older, they not only constitute a growing challenge to the reproductive success of the adults, they also become more able to survive independently and to disperse successfully, thereby diminishing the cost of aggression and ultimately contributing to the reproductive success of their increasingly intolerant parents.

Because parents are equally related to all of their offspring, regardless of the latter's age, it may seem paradoxical that parents discriminate among these offspring, behaving less tolerantly toward those that are older. However, such a pattern is consistent with the material presented above concerning reproductive competition, since older individuals offer more competition than younger ones. It also accords with the expectations of parent/offspring conflict theory (Trivers, 1974), which emphasizes the differing interests of parents and offspring, depending among other things upon the age of the offspring and, therefore, the neediness of those offspring and the likelihood that additional parental investment will rebound ultimately to the parents' benefit or detriment. All other things being equal, we also expect that parents will preferentially invest in younger—and needier—offspring, especially when older offspring constitute a threat to the parents' immediate reproductive success, whereas younger offspring do not. (Note that maturing individuals that do not receive a full measure of parental benevolence are not necessarily thereby deprived of success; rather, they seem more likely to disperse, as a result of which some will be successful, although presumably most will not.) On the other hand, since yearlings and two-year-olds have a higher reproductive value than do young of the year, we might expect some counteracting tendency for females to invest preferentially in the former.

The marmots' primary female-oriented social unit appears to be the female with her offspring, whereas the primary male-oriented social unit is a harem, consisting of one or more female kin groups (Armitage, 1981b; Michener, 1983). Males compete with other males including their relatives—even their sons—for as many female units as possible, while females compete especially with individuals from other such units. It therefore seems likely that kin selection is very much involved in the evolution of marmot sociality, although its intensity seems to be limited by the sometimes competing demands of individual fitness maximization via proximate reproductive success, which sometimes induces individuals to behave nonaltruistically—indeed, aggressively—toward even their own offspring.

Finally, kin selection may also be involved in the evolution of differing reproductive rates among different marmot species. Hoary and Olympic marmots seem to experience less gene flow than do yellow-bellied marmots, which in turn apparently experience less gene flow than do woodchucks (see Chapter 13). As a result, hoary and Olympic marmots seem likely to be viscous populations with a relatively higher average coefficient of genetic relationship than characterizes yellow-bellied marmots, which in turn are more likely to be inbred than woodchucks. If some degree of density-dependent competition occurs in marmot meadows, then selection can be expected to have favored progressively slower reproductive rates in the sequence *M. monax, flaviventris, olympus/caligata.*

A trend of this sort does in fact occur, although it remains doubtful whether it is due to the operation of kin selection *per se.* "Altruistic" reproductive restraint might conceivably be responsible for the observed reduction in breeding rates from *M. monax* to *flaviventris* to *caligata.* Ultimately, altruistic behavior is of particular interest to biologists insofar as it translates into reproductive restraint: whether food-sharing, alarm-calling, or "helping at the nest," altruistic behaviors are especially salient if they diminish the personal reproductive success of the individual in question while enhancing that of another. Periodic nonreproduction among marmots could therefore result, at least in theory, from kin selection favoring reproductive restraint when such restraint reduces competition with close relatives, thereby enhancing the reproductive success of these relatives and, ultimately, the inclusive fitness of the nonbreeder.

If valid, this scenario should generate predictable intraspecies patterns of reproductive restraint as well. Specifically, adult female hoary marmots often skip additional years beyond those expected as a result of adherence to a biennial breeding schedule. If nonreproduction is somehow mediated by kin selection, we might predict that such skipping is more likely to occur when the females in question occupy a colony in which there are many close relatives; similarly, skipping should be less likely when fewer close relatives are present and, therefore, kin selection is less likely to inhibit "selfish" reproduction. I examined all known cases in which adult female hoary marmots skipped an additional year, categorizing their colonies by the coefficient of genetic relationship between skipping females and their co-residents. Contrary to the prediction, skipping tended to be associated with co-residents that are more distantly related to the skipping individual; that is, the average r between skipping females and the other colony residents (.17) was significantly lower than that between nonskipping females and their co-residents (.31) (Fig. 10.12).[33]

This result may well reflect an adaptive modification of female reproduction, with reproductive suppression occurring when females are subjected to higher within-colony competition. In turn, this suggests that contrary to the prevalent view that genetic relatedness predisposes females toward altruistic nonreproduction, kinship may well influence social behavior and the competitive environment—and ultimately reproduction as well—such that individuals whose r is low are more likely to

[33] Mann-Whitney U test, $P < .05$

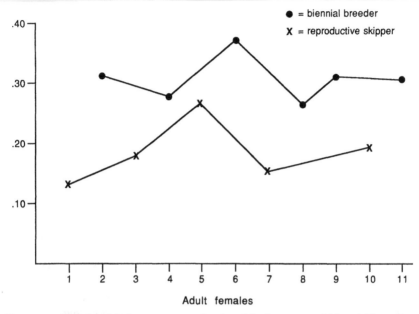

Fig. 10.12. Relationship between reproductive skipping, normal biennial breeding, and genetic relationship between 11 focal adult female hoary marmots (*M. caligata*) and their co-residents. Each number on the abscissa represents a different, known individual.

inhibit their own reproduction, presumably because they experience more within-colony competition. The direct effect of kinship on reproduction *per se* appears to be that increased within-colony relatedness leads to a more amicable, less competitive social environment, in which would-be reproductive females are more likely to prosper. It therefore seems likely that higher within-colony coefficients of genetic relatedness lead to a *greater* probability of reproducing rather than altruistic reproductive restraint. (Of course, this would occur only up to a point; mouths to feed are a drain on remaining resources, whether the individuals in question are related or not.)

Yellow-bellied marmots may show a similar pattern. Female fecundity decreases following male turnovers, when a new and typically unrelated male takes over a colony (Armitage and Downhower, 1974). Such takeovers were too rare among hoary marmots for comparable data to be obtained, although the following cases are suggestive. Two colony males died during hibernation and two were experimentally removed in the late summer. In all four situations, the males were replaced by new adult males, two of which had previously been satellites in the same colonies. Of the three potentially reproductive adult females paired with a new male, not one produced a litter.

This speculation suggests that immigrant females, surrounded by distant relatives and/or nonrelatives, should be more likely to skip than "birthright" colony residents, which are more firmly nestled in a network of kin, thus more likely to experience social benevolence, and, accordingly, more likely to reproduce up to their potential.

Adult woodchucks do not occupy colonies. Hence, the boundaries of woodchuck social systems are not clearly demarcated. Individuals still interact, if only rarely, with their neighbors; such interactions are more difficult to monitor than those among the colonial species, but they may be no less influential with regard to reproductive success. Since virtually all adult female woodchucks reproduce, whereas only a fraction of yearlings do so, it seems reasonable that the latter's "decision" of whether or not to reproduce is sensitive to their competitive (i.e., kinship) environment. Therefore, I arbitrarily defined a measure of "kinship density" for a sample of ten yearling females by recording the average coefficient of relationship between each such female and her neighbors, defined as the yearlings and/or adults within a 100-meter radius of each target yearling. The sample was limited to individuals for which genetic relationships of at least two-thirds of their neighbors were known—in all cases, these individuals were either sibs or offspring. Each female was therefore characterized by the average coefficient of genetic relationship between her and all other known individuals resident within the surrounding 31,400 square meters.

Densities varied between three and six individuals (mean = 3.3), and the average density for skipping females (3.0) was not significantly different than that for nonskipping females (3.8).[34] Of the ten female yearlings whose kinship densities were thus characterized, three reproduced and seven did not. Among the former, the average kinship density was 0.24; among the latter, it was 0.11. Although the numbers are too small for statistical testing, these admittedly limited data suggest that individuals in an environment of high "kinship density" (i.e., high average r with their neighbors) are less likely to skip a reproductive year than are individuals with lower average r. Apparently, therefore, even among the relatively solitary woodchucks, reproduction may be influenced by the coefficient of genetic relationship experienced by individual females.

[34] Mann–Whitney U test, P > .10

In an interesting contrast, a very different set of correlations has been suggested to influence vole (*Microtus*) population dynamics (Charnov and Finerty, 1980). It has been argued that in *Microtus*, low population density leads to *high* coefficient of genetic relatedness, which in turn generates a low rate of neighbor/neighbor aggression. With increasing population density, average r declines and, accordingly, aggressiveness increases. In theory, temporal fluctuations of this sort could characterize longitudinal variations within marmot populations as well; however, comparisons between different species of *Marmota* suggest strongly that the most solitary species (*monax*) experiences a lower effective density— measured by number of social interactions—regardless of the actual population size; moreover, the colonial forms seem to have a higher kinship density.

Although the correlation between kinship density and reproductive success appears to hold interspecifically among marmots, it also is somewhat paradoxical. Thus, if we compare *M. monax* with *M. flaviventris*, and *M. flaviventris* in turn with *M. caligata*, we find a likely increase in kinship density. And yet, we also find a trend of progressively lower reproductive rates. Within a given species a neighborly network of close kin appears to favor reproduction, whereas between different species the factors responsible for reproductive rates are not similarly influenced by kinship density. The same factors selecting for lower reproductive rates in certain species of marmots also appear to select for a social pattern in which genetic relatives tend to be in close spatial proximity.

As we shall see, the differing patterns of reproductive performance among different marmot species seem most parsimoniously attributable to natural selection operating simply upon individuals, each of which is selected to maximize its personal, lifetime reproductive success. However, there is no reason why such factors need be mutually exclusive, and it remains possible that kin selection has operated upon reproductive patterns themselves, just as it has apparently had subtle but real influence on social interactions.

Summary and Conclusions

1. A simple field experiment that established young/young, young/ yearling, and yearling/yearling dyads of known genetic relationship did not reveal any consistent differences in social behavior as a function of genetic relationship, with the sole exception of full sibs, which behaved more affiliatively than did other dyads.

2. Hoary and Olympic marmots appear to experience a low level of gene flow and a high level of genetic relatedness within their colonies; yellow-bellied marmots experience more gene flow; and woodchucks experience the most gene flow and, presumably, the lowest degree of relatedness with their neighbors.

3. Among the marmots, a high level of gene flow tends to correlate with encountering a high proportion of distantly related or unrelated individuals and, in turn, higher social intolerance; cause and effect, however, are difficult to distinguish.

4. Although alarm calling is generally consistent with the expectations of inclusive fitness theory, field evidence suggests that the cost of calling, if any, is slight.

5. Patterns of foraging and the social use of space accord with expectations based on kin selection; in particular, burrowmates tend to be close relatives.

6. Among two-year-old and older animals, kinship patterns influence much of hoary marmot social behavior, with the exception of the adult males, for which male/male competition overrides kinship effects.

7. Burrow sharing is both a consequence of kinship and also a means by which genetic relatives achieve familiarity, which in turn correlates with reduced competitiveness.

8. Among the high-elevation species, intracolony interactions tend to be amicable, whereas intercolony interactions tend to be competitive.

9. Especially among yellow-bellied marmots, kin-biased behavior seems to effect social interactions; colonies develop as a system of bifurcating matrilines with competitiveness prevailing between such lines and tolerance prevailing within them.

10. Among all marmots, competitiveness increases with increasing age of the interactants—although, of course, genetic relatedness remains unchanged; species maturing earlier develop intolerance earlier.

11. The primary social unit for males appears to be the harem (one or more female kin groups), whereas for females it is her offspring; the pressures of reproductive success, acting differently on males and females, limit the expression of kin selection.

12. Higher within-colony genetic relatedness, rather than leading to altruistic reproductive restraint, apparently leads to a higher frequency of reproduction, because the competitive environment is more relaxed in such cases.

PART

III

Population Biology

Death feeds us up, keeps an eye on our weight,
and herds us like pigs through the abattoir gate.
                              Palladius (4th-5th century)

Every moment Nature starts on the longest
journey, and every moment she reaches her
goal.   Goethe

Just as the student of behavior observes the *doings* of individuals, the population biologist is concerned with their comings and goings. And here, in the numerology of births and deaths, is the bottom line of the evolutionary calculus. Successful genes, individuals, or populations are those whose numbers ultimately maintain themselves, and the failures are those for whom deaths exceeds births. The successful—genes, individuals, populations—are those whose behavior (and anatomy, physiology, etc.) has adapted to environmental circumstances. The failures have done less well. Moreover, by looking at the success or failure of different behavioral tactics, as measured by their population outcomes, we can interpret and analyze the evolutionary payoffs and compare them with alternative tactics—which presumably is more or less what natural selection has also been up to. In Part III, therefore, we shall examine those behaviors that relate most directly to fitness, and to evolutionary success and failure.

CHAPTER II

Mortality and Survivorship

For the student of population biology, death is problematic, life, less so. In some cases, if we are lucky, we can actually observe the death. Generally, however, when an individual disappears we cannot know the cause of death with certainty, and indeed, we cannot even know for certain that death has occurred.

It is also problematic whether death by human causes—notably, automobiles or hunting—should be included in mortality calculations for free-living marmots. Such events do not appear to be "natural." But they are quite real. Hunters are significant predators on marmots of all species; automobiles, on many. Not surprisingly, therefore, most research on these animals is conducted away from roads or in reserves, so that obvious human-caused mortality is only rarely at issue. The problem of assessing cause of death, or even the fact of death, however, remains. Hunting pressure is particularly important among many woodchuck and certain low-elevation yellow-bellied marmot populations. In its absence, the major causes of marmot mortality appear to be overwintering deaths, especially of the hibernating young, and dispersal, which leads ultimately to death via predation, exposure, starvation or, alternatively, nonreproduction through failure to obtain a mate in a viable environment. It should be noted, however, that Kizilov and Berendaev (1978) maintained that for the long-tailed marmot the highest mortality occurs within two weeks after spring emergence.

Marmots are clearly vulnerable to predation, and, as we have seen, their behavior seems influenced by the risk of predation. It nonetheless appears that predation of colony residents is less of a factor than over-winter mortality or dispersal; among woodchucks, however, predation alone may be comparable to the other two sources of mortality. In assessments of potential overwinter mortality, it is generally assumed that animals last seen in late summer attempted to hibernate, so that any failure to emerge the following spring is due to mortality at some point during the winter.

Only about 49% of hibernating young yellow-bellied marmots are recaptured as yearlings (Armitage and Downhower, 1974), although there is considerable year-to-year variability. A similar pattern occurs among high-elevation populations of the same species, in which, for example, only 15 of 32 overwintering young emerged (Armitage and Johns, 1982). Substantial variation also occurs in the age structure of populations of *M. bobak* (Shubin, Abelentsev, and Semikhatova, 1978), with young of the year constituting 69% of the total population one year and 21% in another. This fluctuation was due to variation in mortality of the young.

If young animals simply starved in their hibernacula, then we would expect a continuum of relative malnourishment to be apparent among those yearlings that survive. However, yearlings that emerge in the spring rarely seem emaciated, so starvation alone seems unlikely. Perhaps general physical debility, associated with malnourishment, establishes a threshold of vitality under which individuals succumb to hypothermia or lethal disease. Free-living woodchucks, for example, suffer from pulmonary granulomata and other lesions, commonly associated with bacteria of the genus *Mycobacterium*, fungi, and foreign bodies, which can result in possibly lethal pneumonias (Snyder, 1976). Significantly, a high incidence of bronchopneumonia is characteristic of natural populations of *M. monax*, especially in late winter or early spring; it may well be that the hibernating environment, with its limited ventilation, provides an especially suitable medium for such ailments, which may well be lethal in underweight, malnourished individuals. There seems no reason to expect that woodchucks are unique among marmots in this respect, although epidemiologic studies have not yet been conducted among any other New World species of *Marmota*. Free-living woodchucks also suffer from glomerulonephritis and liver cancer and are known carriers of the hepatitis B virus (Wright, Tennant, and May, 1987).

The role of epidemics is unclear among the Old World species, although plague has been documented in the tarbagan (Letoff, reference in Zimina, 1978) and suggested for the bobak (Shubin et al., in Zimina, 1978) and gray marmots (*M. baibacina*) (Bubikoff and Berendaev, in Zimina, 1978). The Old World species are also exploited for fur and, to a lesser extent, for food. Mortality among black-capped marmots appears to be especially associated with hibernation and following dispersal.

Permafrost makes dispersal especially hazardous to this species, because it restricts opportunities for establishing new, viable burrows (Kapitonov, 1960).

Whatever the precise cause of death over winter, nutritional status appears to be implicated. Thus, 58% of young yellow-bellied marmots that first appeared above ground before July 15, as opposed to only 19% of young that appeared above ground after July 15, were recaptured as yearlings. The former tend to weigh 1.4–1.8 kg by mid-September, whereas the latter average about 1.16 kg (Armitage and Downhower, 1974). It seems that lighter young—in this case, those whose weight gain is delayed because of later emergence—are less likely to survive hibernation. The percentage overwinter survival of young yellow-bellied marmots in Colorado is highest for those animals weaned during the second week of July and becomes progressively lower for litters weaned either earlier or later (although it is skewed toward the former). Litter size, in turn, is largest for early litters, declining with season (Fig. 11.1).

The number of yearlings produced per litter, prior to dispersal, is the product of litter size and percentage survival: this reveals a pattern similar to that of percentage survival: weaning during the second week of July results in the largest number of surviving yearlings, once again with the distribution skewed toward the early dates (Fig. 11.2). And finally, the frequency of litters actually weaned at different dates corresponds closely to the optimum seasonal scheduling, suggesting that as with many other animals, yellow-bellied marmot reproduction corresponds with those times that lead to maximum success, although with some variability. More litters of this species are weaned during the second week of July than at any other time, with the distribution skewed toward the early dates (Fig. 11.2). The frequency distribution of yellow-bellied marmot weaning dates corresponds very closely to the distribution of overwinter survivorship.

Weaning dates among hoary marmots are more seasonally compressed, occuring during a 3-week period between mid-July and early August, as opposed to the nearly 8-week span of *M. flaviventris*. This reflects the shorter above-ground season of *M. caligata*, due especially to the lateness of the spring snowmelt in its range. Within this narrower "window of weaning," however, *M. caligata* shows a pattern very similar to that of *M. flaviventris*: percentage survival is highest among mid-range weaners (the last week of August; Fig. 11.1), litter size declines with season (Fig. 11.2), the number of yearlings produced per litter is highest in mid-season (Fig. 11.2), and most litters are weaned during mid-season—the last week of July—when survivorship is highest.

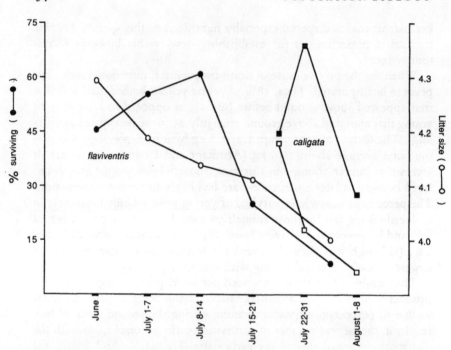

Fig. 11.1. Litter size and percentage of young surviving their first hibernation as a function of date of weaning among *M. flaviventris* and *M. caligata*. *M. flaviventris* data from Andersen, Armitage, and Hoffmann (1976). Final data points for *M. flaviventris* include all litters weaned during the last week of July and later.

Early-emerging young hoary marmots tend to be the offspring of dominant females: 53% of young emerging before July 26 had mothers whose chase ratios exceeded 1.5, whereas only 11% of those emerging after July 26 had mothers with chase ratios that high.[1] Although it is possible that dominant females wean their offspring more quickly than subordinates, it seems most likely that dominant females simply breed earlier. Data are available on the dominance status and seasonal emergence dates for 12 adult female hoary marmots: the two factors are positively correlated, although not strongly.[2] It may be that dominant females are better nourished and hence better able to emerge early, profiting from this by ultimately producing offspring with a higher survivorship. Alternatively, dominant females may somehow delay the emergence of their

[1] binomial test, P < .05
[2] Spearman rank correlation, .10 > P > .05

less dominant colleagues, perhaps because the latter are more fit breeding somewhat later, thereby diminishing their personal competition with other females and reducing possible competition among their litters.

The basic lifetime mortality and survivorship pattern for Olympic and hoary marmots generally parallels that of yellow-bellied marmots, except that mortality factors for *M. caligata* tend to be extended by about one year (Fig. 11.3). Among Olympic marmots, 40 young of the year produced 11 dispersing two-year-olds two years later (Barash, 1973a). The hoary marmot's overwinter mortality—although high—is less than that of the yellow-bellied marmot, and its lifespan appears to be greater, although data are insufficient at present. Such a pattern would be consistent with the larger body size and somewhat more *K*-selected life-history pattern of hoary marmots (see Chapter 12). Among woodchucks, the average mortality rates are 77% for the young and 30% for adults (Snyder, 1976). Given its relatively *r*-selected life history, this species should exag-

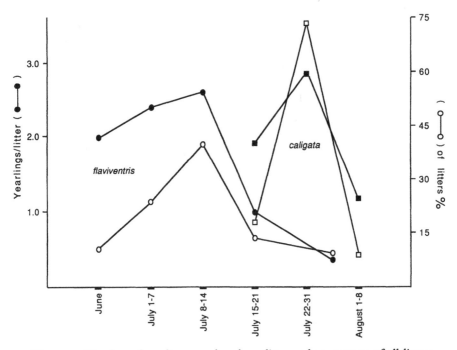

Fig. 11.2. Number of yearlings produced per litter and percentage of all litters produced, as a function of date of weaning among *M. flaviventris* and *M. caligata*. (*M. flaviventris* data as for Fig. 11.1.)

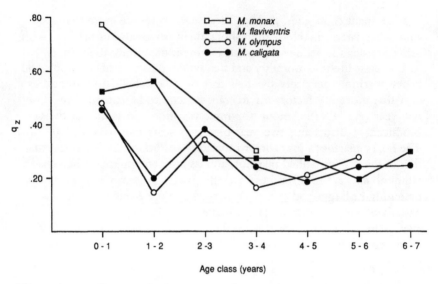

Fig. 11.3. Mortality rates for four marmot species. (Data for *M. flaviventris* from Armitage and Downhower, 1974; for *M. monax*, from Snyder, 1976. Note that age data beyond the first year do not exist for *monax*, and therefore mortality rate for "adults" has been arbitrarily placed at 3–4 years.)

gerate the *M. flaviventris*-like mortality patterns, with shorter average lifespan and relatively early mortality factors. Given that young woodchucks disperse and dispersers are almost certainly more susceptible to predation and social stress, and are also likely to have suboptimal foraging opportunities, time budgets, and refuge options, *pre*-hibernation mortality of young woodchucks might well be a significant addition to the patterns of overwintering mortality that are apparently experienced by all marmots. I noted 47 road-kills of woodchucks near Oneonta, New York, over three years; of these, 39 were young of the year and only eight were yearlings or adults. This disparity may reflect inexperience on the part of young animals, as well as the fact that dispersing individuals are likely to be wandering greater distances and through terrain with which they are unfamiliar.

Data on overwinter mortality of individual woodchuck young of the year have proven difficult to obtain because essentially all individuals disperse and dispersers are difficult to monitor. Mark/recapture data, however, suggest that about 30% of adult woodchucks and about 50% of young of the year die annually (Snyder, 1976). By observing a small number of young woodchucks that did not disperse far (generally less than 300 meters), I was able to gather some suggestive data: of 32 *M. monax* young of the year, I observed only six as yearlings in a succeeding year. Given the shorter hibernation and longer above-ground period of *M. monax*, one might predict that the effect of the date of weaning on overwinter mortality is less significant for that species than for *M. flaviventris, olympus,* or *caligata*. However, of the 32 young woodchucks described above, 19 were weaned before June 8 and 13 were weaned after; all six of the surviving yearlings were in the former category.[3]

But even these limited data do not necessarily indicate a clear correlation for *M. monax* between date of weaning and mortality during the first year. Thus, as we have already seen, some woodchucks reproduce as yearlings, and yearlings emerge from hibernation later than adults. Hence, they wean their young later as well. Higher mortality among late-weaned young could reflect higher mortality among the offspring of yearlings, and this is in fact the case. Of the 32 young of the year discussed above, nine were produced by yearlings: all of these were weaned after June 8, and none of them apparently survived to become yearlings the following spring.

In *M. flaviventris*, there is no difference between the survivorship of male and female young of the year: 51% vs. 48%, respectively, at the high elevation of 3,400 meters (Johns and Armitage, 1979) and 47% for both sexes at medium elevation (K. B. Armitage, personal communication). The same is true for *M. olympus* (55% and 52%) and *caligata* (56% and 54%). By contrast, however, survivorship of yearling yellow-bellied marmots is sex-biased, 24% for males vs. 64% for females at high elevation; for adults, the comparable figures are males 53% and females 90% (Johns and Armitage, 1979). At medium elevation (2,900 meters), the survivorship figures for yearlings are 46% for females and 6% for males, although this includes losses due to dispersal, which, as we have seen, is strongly sex-biased. Clearly, some dispersers are successful, so these data are overestimates of mortality.

By contrast, survivorship of male and female yearlings is not distinguishable for either *M. olympus* (84% and 86%) or *M. caligata* (88% and 83%). However, survivorship of two-year-olds of *M. caligata* is biased definitely toward females (61% vs. 85%); insufficient data are available for *M. olympus*. The trend thus appears to be that among those species maturing later, male/female differences in survivorship are apparent

---

[3] Fisher's exact test, P < .05

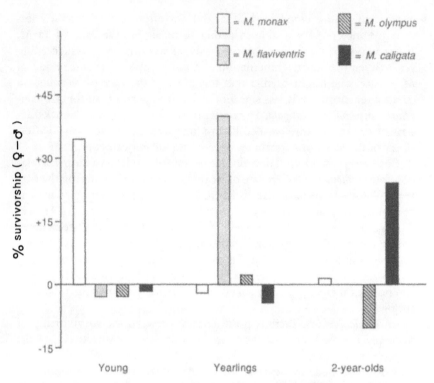

Fig. 11.4. Differences between female and male survivorship among four species of marmots, as a function of age. (Data on *M. flaviventris* from Johns and Armitage, 1979; note also that data for *M. monax* presented as "two-year-olds" actually apply to "all adults.")

later. (It should also be noted that for each species the difference between the sexes' mortalities coincides with age at dispersal; see Chapter 13.)

Unfortunately, similar data are once again difficult to gather for *M. monax*; however, of 53 dispersing young woodchucks whose sex was known (28 males and 25 females), I located 26 in late summer or early fall. Eighteen were females (34% of the original cohort) and eight were males (15%). The following spring, 11 females (44% of the original cohort) and only three males (10%) were identified. This difference could, of course, be due partly to males dispersing farther than females. It also suggests that the male/female difference in mortality occurs earlier in those marmot species in which maturity—i.e., male/female differentiation—occurs earlier (Fig. 11.4).

Weather conditions also influence mortality, at least among the montane species. In *M. flaviventris*, survival is higher when winter ends earlier, although the duration of winter *per se* has little relationship to survivorship (Armitage and Downhower, 1974). Reproductive females are a prominent exception: for them, higher survivorship correlates with a later onset of winter (Armitage and Downhower, 1974). Reproductive female marmots begin gaining weight significantly later than do their nonbreeding colleagues, and, accordingly, during years of early snow accumulation, females who have borne young may be unable to gain enough weight to survive hibernation.

Drought appears to be a major determinant of mortality for the bobak or steppe marmots. However, drought conditions during the preceding year had no observable effect on survival of young, perhaps because of a homeostatic system: in good years, animals hibernate earlier (before drought becomes a problem), whereas in bad years, they remain active longer into the late summer and early fall. The crucial factor determining mortality or survival for young of the year appears to be the amount of forage available during their mother's lactation (Shubin, Abelentsev, and Semikhatova, 1978).

Among Olympic and hoary marmots, winter mortality among young animals, but not adults, correlates inversely with snow cover—relatively few animals, young or adult, die during winters of heavy snow, whereas hibernating young of the year are especially subject to mortality when the snow cover is sparse (Fig. 11.5). The Olympic and Cascade Mountains, with their maritime climates, experience highly variable winters, with heavy winds and sudden January rains not uncommonly leaving slopes temporarily bare during the coldest time of the year. Because of the insulating qualities of snow, it is not surprising that the absence of snow cover in midwinter leads to high mortality, particularly among hibernating young whose small body size renders them especially susceptible to thermal stress, which almost certainly exacerbates mortality from disease, hypothermia, or starvation.

Among both Olympic and hoary marmots, annual mortality influences dispersal: when overwinter mortality is high and therefore the colony is small, two-year-olds are less likely to disperse than when overwinter mortality is low and the colony's population is accordingly high (Fig. 11.6). It might be supposed that environmental conditions yielding high overwinter mortality could also produce weaker two-year-olds, which, because of their relatively poor condition, do not disperse. In this case, nondispersal of two-year-olds would be due primarily to their poor

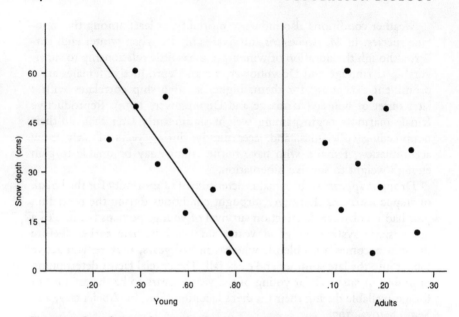

Fig. 11.5. Correlation of snow depth with winter mortality of hoary marmot (*M. caligata*) young and adults. Snow depth is presented as the average depth at three marmot habitats for the three months between December and April at which snow cover was lowest during each of ten years, 1974–83; data were provided by the Northwest Avalanche Center, National Oceanographic and Atmospheric Administration, Seattle, Washington. Mortality records for each data point represent a minimum of two colonies and a total of 32 marmot colony-years.

physical condition and be correlated only incidentally with colony population density. However, the weights of dispersing and nondispersing two-year-old Olympic and hoary marmots are not statistically distinguishable in a sample of 11 nondispersing and 12 dispersing hoary marmots and nine dispersing and eight nondispersing Olympic marmots.[4]

The situation is less clear among yellow-bellied marmots: although the number of young and the number of yearlings tend to be inversely correlated, the relationship is not statistically significant (Armitage and Downhower, 1974). The recruitment of yearlings seems more closely tied to the number of resident adult females (and to their individual characteristics: Armitage, 1984) than to the number of young produced per female or to overwinter survivorship of those young. In general, density-

[4]Mann-Whitney U tests, p >.20

dependent effects are not pronounced among yellow-bellied marmots (Armitage, 1975, 1977) with regard to either the maintenance of colony population size or the consequences of colony size for the behavior of colony members.

It seems that colony population size is maintained with more homeostatic precision among the high-elevation hoary and Olympic marmots than among the medium-elevation yellow-bellied marmots. The determinants of local population size among woodchucks are little known, although density independence is probably even more prominent than among yellow-bellied marmots. On the other hand, experimental removal of adult and yearling woodchucks from an intensively studied

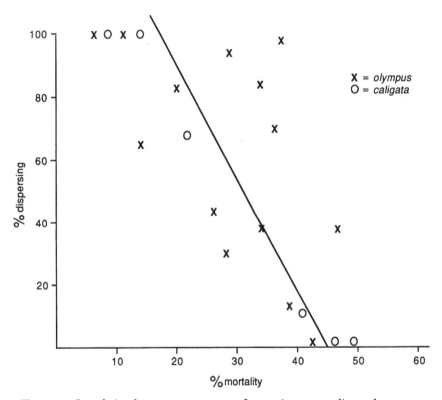

Fig. 11.6. Correlation between percentage of overwinter mortality and percentage of surviving two-year-olds that disperse, among both *M. olympus* and *M. caligata*.

free-living population in central Pennsylvania caused a significant increase in the overwinter survivorship of young animals, leaving their mortality only slightly higher than that of adults and significantly lower than is usually observed among young in a nonexploited population of the same species (Snyder, 1976). This suggests a possible role of density dependence for *M. monax*: mortality may be less a function of the small body size of dispersing young than of the fact that they are dispersing and therefore competing with already established residents. Such competition seems likely to be especially intense over suitable hibernacula, since without an adequate site for hibernation, mortality is a virtual certainty. But whatever the relative importance of the various factors, it seems clear that dispersal, presumably combined with the dispersers' small size and, possibly, their lack of experience, is likely to produce high mortality.

Although the fate of dispersing woodchucks appears to be at least partly density-dependent, the initial occurrence of dispersal seems to be independent of local population density. Because woodchuck reproduction is annual and the young disperse during their first year, there appears to be even less opportunity—as well as less need—for recruitment to be adjusted to mortality patterns than among the montane species. It therefore seems likely that another general trend exists, with *M. monax* reproducing and dispersing with least regard to the immediate social milieu, *M. flaviventris* being somewhat responsive, but not very much so, and finally, *M. olympus* and *caligata* being the most finely tuned.

Such a pattern is consistent with the higher level of average genetic relationship experienced, in turn, by *M. monax*, *flaviventris*, and finally *olympus/caligata*, a sequence that is linked to a progressively denser pattern of social interaction, larger body size, and longer lifespan. Although it may be tempting to interpret this system as reflecting progressively stronger pressures for the maintenance of optimum population size, such an interpretation implies group-level (or rather colony-level) adaptations. Such an interpretation is tenable, especially since the higher montane species also tend to be those in which genetic relatedness within local breeding groups is highest and in which gene exchange between such groups is lowest, thereby approaching the conditions necessary for group selection to operate. However, the pattern is also compatible with selection acting at the level of individuals and their constituent genes: given that male/male and female/female competition reduces the likely reproductive success of nondispersing individuals within locally dense colony populations compared to that of sparser populations, the cost/benefit equation could well tilt in favor of dispersal from the former and nondispersal from the latter, entirely as a result of selfish, individual fitness maximization.

Moreover, inclusive fitness considerations could bestow a further evolutionary payoff on individuals that minimize competition with their kin by dispersing when overwinter mortality is low and springtime population size is consequently high.

Detailed study of the demography of *M. flaviventris* has also revealed substantial differences between colony residents and peripherals (Armitage and Downhower, 1974). Although peripheral females produce litters that are as large or larger than colonial females' (4.46 vs. 4.15), the former are more likely to skip a reproductive year, either because of more frequent nonovulation or because of some mortality factor acting prior to emergence of young. As a result, peripheral females produce an average of 1.14 young and colonial females produce an average of 1.99 young. Furthermore, no yearlings have been attributed to peripheral females, so their actual fitness may well be very low (Armitage and Downhower, 1974; Svendsen, 1974). In *M. flaviventris*, the higher survivorship of offspring from colonies can be attributed to many factors, including the benefits of group living (such as predator avoidance), better forage quality, better hibernacula, greater attentiveness from their mothers, and/or the presence of a resident adult male, who prevents disruption by transient males.

In a sense, all woodchucks are "peripheral," whereas only a limited number of yellow-bellied marmots appear to be peripheral at any given time. Those that are may switch from peripheral to resident when the opportunity permits (K. B. Armitage, personal communication). By contrast, no adult female hoary or Olympic marmots seem to fit this designation. (It is also possible, however, that peripheral individuals are more frequent and a more significant aspect of marmot biology than is generally realized, since group-living individuals are much more noticeable.)

The average age at death is 53 months for captive woodchucks and 12 months for free-living animals. The oldest reported free-living individual lived to an age of 78 months (Snyder, 1976). By contrast, reproducing female yellow-bellied marmots have been reported in their tenth year (Armitage, 1984), and two individuals have been documented to survive to age 13 and one to 14, although none of these reproduced (K. B. Armitage, personal communication). It appears rare for yellow-bellied marmots to survive beyond their eighth year. I have records of reproducing nine-year-old hoary marmots and the life span in this species may well prove to exceed that of *M. flaviventris*; certainly, it is longer than that of *M. monax*.

Summary and Conclusions

1. Predation is an unpredictable cause of marmot mortality, difficult to document but nonetheless probably important; among colony residents, it does not appear to be age- or size-specific. Winter mortality, during hibernation, is more predictable and clearly important; although the precise cause of death is typically unknown, body size appears relevant: smaller animals are more likely to die. Dispersing individuals seem likely to be especially vulnerable to various sources of mortality, although it is very difficult to document this.

2. Overwinter mortality is highest among young of the year and is related to foraging opportunities via date of weaning; young animals that are weaned too early or too late are less likely to survive. The peak in weaning date corresponds closely to that generating maximum overwinter survival. The relationship between emergence and survivorship is likely to be important but is currently little understood.

3. The mortality of males is somewhat higher than that of females, with this difference appearing at different ages in the different species as a function of age at sexual maturation.

4. Although overwinter mortality is not strongly density dependent, dispersal among Olympic and hoary marmots is influenced by mortality among hibernating young and thereby exerts a modest homeostatic influence on colony size.

5. Such fine tuning is least developed among the least social, more rapidly maturing, and earlier dispersing species; this difference seems attributable to selection acting at the level of individuals and their genes rather than at the level of groups.

Reproductive Effort

Reproductive effort is the proportion of its total energy that an organism devotes to reproduction, as well as the patterning of that reproduction. Natural selection operates on the reproductive effort expended at any given time so as to maximize the reproductive value of an individual at that time: the number of successful offspring likely to be produced in its lifetime. Reproductive effort contributes positively to an organism's fitness, but at a cost, measured as a reduction in the individual's remaining reproductive value. If reproductive efforts cause exhaustion, metabolic stress, or increased risk of predation, then, as a result, the expender of such effort is somewhat less likely to reproduce successfully again in the future. Furthermore, we assume that selection maximizes *lifetime* reproductive value and fitness per reproductive effort. This may require submaximal effort at a given time, even though we can assume that ultimately living things are devoted entirely to optimization of their reproductive effort.

Matters of Weight and Age

Perhaps the simplest index of reproductive effort is the weight of offspring produced, i.e., the weight of an average offspring multiplied by the mean litter size. By definition, then, large animals such as elephants make an enormous reproductive effort, whereas the reproductive effort of a mouse is minuscule. A more useful index of reproductive effort is therefore relative weight-specific reproductive effort, calculated as the ratio of total offspring weight to maternal weight. For practical purposes, it is convenient—as well as biologically meaningful—to take maternal weight as the minimum normally attained by adult females: for marmots, the weight of parous females shortly after spring emergence.

The ratio of weight at weaning to minimum adult female weight provides a convenient index of the amount of reproductive effort invested in the production of a single offspring, and similarly, the ratio of litter weight to minimum adult female weight provides an index of the total reproductive effort invested in a given bout of reproduction. Relative weight-specific reproductive efforts for marmots are as follows, based on the weaning weight of female young: *M. monax*—0.81 (calculated from Snyder, Davis, and Christian, 1961); *M. flaviventris*—0.82 (from Armitage, 1981b); *M. olympus*—1.38; and *M. caligata*—1.57.

From these results, it seems that woodchucks and yellow-bellied marmots are virtually identical in their reproductive efforts, whereas Olympic and hoary marmots are also very similar to each other but expend approximately double the effort of the foregoing two species. However, this impression is misleading. During any one year, a female either does or does not reproduce, and when she does, if she is a hoary or Olympic marmot, she invests nearly twice as much metabolic material in her offspring, relative to her own body weight, as does a female yellow-bellied marmot or woodchuck. To compare the reproductive efforts of different marmot species, it is helpful to have an index that incorporates *lifetime* reproductive effort; since some species routinely skip a reproductive year, such a measure must consider the frequency of actual reproductions.

We can assume that essentially 100% of adult female woodchucks are pregnant annually (Snyder and Christian, 1960). By contrast, Olympic and hoary marmots reproduce biennially at best, never breeding more frequently than once every two years, and yellow-bellied marmots often skip a year, though not as regularly as the former two species. Hence, an estimate of lifetime relative weight-specific reproductive effort can be generated by multiplying the former figures by the percentage of adult females actually reproducing during any year: *M. caligata*—24%; *M. olympus*—21%; and *M. flaviventris*—48% (Armitage, 1986b). The appropriate indices so generated: *M. monax*—0.81; *M. flaviventris*—0.39; *M. caligata*—0.30; *M. olympus*—0.28. Few comparable data are available for the Palearctic species, but Nekipelov (1978) reported that 17–77% of adult female tarbagans breed in any given year.

The larger-bodied, slower-breeding montane species make the smallest relative reproductive effort, *M. flaviventris* makes more, and *M. monax* makes the most. Perhaps another, even better index of reproductive effort could be obtained by multiplying the relative weight-specific index by the average number of litters produced by a female of each species during her lifetime. In addition to including species-specific differences in reproductive rates, such an index would also be sensitive to possible differences among the different species in lifespan and age of sexual maturity. Unfortunately, such data are not yet available. It may also be misleading to extrapolate the energetic costs of reproducing across different habitats and social systems. It is not obvious, for example, that producing an offspring of a given weight—or even of a given weight relative to total adult weight—is equally costly for individuals inhabiting colder montane environments and for others inhabiting warmer lowland environments.

It is nonetheless tempting to interpret the above findings as reflecting an *r*-to-*K* continuum, with woodchucks being the most *r*-selected, hoary and Olympic marmots the most *K*-selected, and yellow-bellied marmots roughly intermediate. This is consistent with the lower reproductive rate, later maturity, and more integrated social systems of *M. olympus* and *caligata*, and the consistent trend through *M. flaviventris* to *M. monax*. However, marmots pose some inconsistencies for such an analysis (Armitage, 1981b). For example, (1) the larger and presumably *K*-selected species produce relatively smaller young, an *r*-selected trait, and (2) lactation in the montane species takes 25–30 days, whereas among woodchucks, it takes up to 44 days—suggesting that the *latter* is *K*-selected. Looking closely for specific factors selecting for the given reproductive regime in each case therefore seems more profitable than seeking to place each species in its "rightful" place on some hypothetical *r*-to-*K* continuum.

Given that annual body weight fluctuates so greatly among marmots—because of hibernation—and given that prehibernation body weight correlates with mortality, it seems appropriate to search for additional correlations between body weight and reproductive effort. The natural-history facts already quoted in this respect suggest a coherent pattern. Woodchucks typically lose weight after spring emergence, since forage quality is generally low at this time, and the animals live off their remaining body fat. In northeastern North America, yearling males lose weight for only a few weeks, until the end of March, whereupon they gain throughout the summer. By contrast, yearling females continue losing weight until the end of April, which is consistent with the fact that yearling males do not breed, whereas up to 50% of yearling females do. Adult male woodchucks—which emerge several weeks earlier than the yearlings—don't begin regaining weight until mid-April, about three weeks after the yearling males. As we have seen, this is presumably because adult males are actively reproducing, whereas yearling males are not.

The weight pattern for adult female woodchucks, in turn, resembles that of yearling females: weight gain is delayed until the end of April. It therefore appears that post-emergence weight loss is a component of reproductive effort for both males and females. Among the former, this cost is most readily apparent shortly after spring emergence and is presumably associated with the rigors of male/male competition. For females, it occurs notably via delays in their ability to begin regaining weight after their pregnancies.

Adult female woodchucks do not use up their fat stores until the end of May; adult males, by contrast, have used up their remaining hibernation fat a full month earlier, by late April (Snyder, Davis, and Christian, 1961). As a result, adult males begin replenishing their fat earlier (early June) whereas adult females do not do so until early July. Nonetheless, there is no evidence that adult females enter hibernation any later than adult males. I trapped and sexed 14 different adult woodchucks during October in central New York State: six were females and eight were males. The lengthy active season presumably provides enough opportunity for females to put on sufficient fat without their having to delay their hibernation.

By contrast, reproductive female hoary marmots (and probably Olympic marmots as well) remain above ground, foraging and replenishing their fat stores significantly longer than adult males, yearlings, or two-year-olds (Barash, 1976b; Fig. 12.1). Among these species, the above-ground active season is brief, and, accordingly, adult females emerge from hibernation only about two weeks later than the adult males. Presumably, they need as much time as possible to gain weight for the next hibernation, and, moreover, they cannot emerge much later than the males if their offspring are to have sufficient opportunity to gain weight themselves. It seems likely, therefore, that fat stores in male and female Olympic and hoary marmots are used up at about the same time; as a result, females must remain active for a longer period to replenish losses incurred because of breeding.

Pregnant hoary marmots do not gain weight significantly, even with their growing fetuses. Weight gain does not occur during lactation either, but only after the young are weaned. This further emphasizes a likely role for post-hibernation fat stores, since the reproductive success of females may well depend on whether they emerge from hibernation with sufficient fat to permit pregnancy and lactation; moreover, survival may also be at issue, given the need of reproductive females to gain weight

% FORAGING

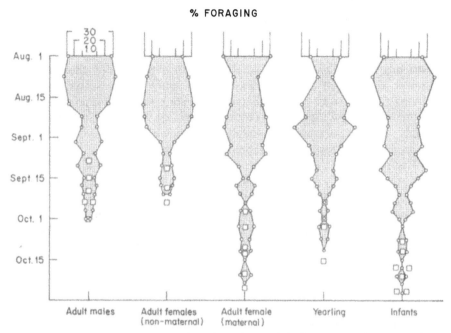

Fig. 12.1. Immergence dates (last day of recorded above-ground activity) for hoary marmots (*M. caligata*), as well as percentage of above-ground time spent foraging. Each square represents an individual; foraging polygons are based on a minimum of 50 point-censuses per animal per day.

after their young are weaned and before hibernating once again. It may be that some underweight females reproduce normally but then have difficulty accumulating enough fat to hibernate successfully.

Similar considerations seem to apply to yellow-bellied marmots: in one reported case, an adult male and a nonreproductive female began hibernating between 11 and 17 September, whereas a reproductive female did not hibernate until 23 September (Kilgore and Armitage, 1978). Reproductive female yellow-bellied marmots gain weight more slowly than nonreproductive females (Armitage, Downhower, and Svendsen, 1976), and only after their young are weaned (Andersen, Armitage, and Hoffmann, 1976). As a result, by late August and early September, reproductive female yellow-bellied and hoary marmots weigh less than their nonreproducing colleagues. Furthermore, in the case of female hoary marmots at least, even their prolonged above-ground foraging is not quite sufficient to close the gap: during the first three weeks following spring emergence, 16 females that reproduced the previous year weighed somewhat less than 14 females that had not reproduced (5.3 kg vs. 5.8 kg; SD = 0.4).[1]

Female hoary marmots that expended reproductive effort by breeding during year 1 do not expend additional effort by breeding once again in year 2; i.e., they breed in alternate years only. Reproductive cycles among the Palearctic species have unfortunately not been reported, although breeding female steppe marmots begin accumulating fat "considerably" later in the season than adult males or nonreproductive adult females (Shubin, Abelentsev, and Semikhatova, 1978).

Because nearly all adult female woodchucks reproduce annually, it is unfortunately not possible to collect a meaningful sample of barren females; if this could be done, however, it would likely show no detectable differences between their weights and those of breeding females by the time of hibernation, since woodchucks not only breed earlier than their montane congeners but typically enter hibernation later, by which time their reproductive weight loss has been replenished.

Among woodchucks, about 33 to 50% of yearling females reproduce (Snyder and Christian, 1960). No yearling yellow-bellied marmots reproduce, and 6 of 23 known two-year-olds reproduced (Armitage and Downhower, 1974). Among Olympic marmots, in which no two-year-olds breed, 10 of 24 known three-year-olds reproduced (Barash, 1973a); similarly, 26 of 60 three-year-old hoary marmots produced a litter. Given the demographic effects and consequent fitness advantages of early breeding, we can expect that selection would favor reproducing as early as possible. It seems reasonable, therefore, that yearling woodchucks, two-year-old yellow-bellied marmots, and three-year-old Olympic and hoary marmots should reproduce if they can do so without prohibitive cost to their lifetime reproductive success.

The inclination to expend reproductive effort varies with the cost as well as the benefit, and given the foregoing correlations between weight and reproductive effort, it seems reasonable that a primary determinant of reproductive cost should be body weight—or, more properly, body weight relative to the minimum adult weight—such that the early breeders in each species (yearling woodchucks, two-year-old yellow-bellied marmots, three-year-old hoary and Olympic marmots) should be the heavier females. Furthermore, we might expect that when it comes to age at first breeding, the montane species should be more weight-sensitive, with *M. monax* the least sensitive of all. In a sense, this is the

[1] t test, P < .10

case, since as we have seen, essentially 100% of adult woodchucks breed, regardless of their body condition. However, with regard to age at first breeding, female woodchucks are probably no less weight-sensitive than females of the other species, since to an undersized yearling woodchuck the cost of breeding is presumably comparable to the cost of breeding experienced by an undersized yellow-bellied marmot two-year-old, or an undersized hoary or Olympic three-year-old.

For *M. caligata*, we can predict that breeding three-year-olds will be, on average, heavier than nonbreeding three-year-olds; i.e., that weight likely determines when to begin breeding. However, any such relationship is confounded by the pattern of female/female competition described earlier, such that a three-year-old hoary marmot, if co-resident with a dominant, older, reproductive female, is unlikely to breed, regardless of her own physical condition. An admittedly small sample of monogamous three-year-old hoary marmots nonetheless suggests a likely pattern: of seven such individuals, three reproduced. These had a median weight at spring emergence of 3.2 kg; the four nonbreeding three-year-olds had a median weight of 2.6 kg. No three-year-old female weighing less than 2.5 kg reproduced.

By contrast, the solitary nature of woodchucks likely makes them less sensitive to female/female competition. As a result, the condition of yearlings should be less confounded as a determinant of reproductive status. I trapped 22 yearling female woodchucks, of which seven subsequently reproduced; these averaged 2.3 kg (SD = 0.23), whereas the 15 nonbreeders averaged 2.0 kgs (SD = 0.18).[2]

It seems reasonable to measure reproductive effort in energetic currency, although in fact convenience is a major reason for its use. Natural selection ultimately operates on total benefits and costs, with both measured in units of inclusive fitness. Although absolute energy costs seem likely to correlate with total reproductive costs, and energy costs relative to total body mass seem likely to correlate even better, the relationship could be misleading, since larger animals, for example, could afford larger relative as well as absolute expenditures in their offspring without reducing their remaining reproductive value. It remains to be seen whether breeding marmots actually have a lower life expectancy. Limited current data do not suggest this: I examined possible correlations between lifespan and age at first reproduction among female hoary marmots whose lifespan and reproductive histories I knew, and found no significant—or even consistent—relationship.[3]

[2] Mann-Whitney U test, P < .10
[3] Spearman rank coefficient, P > .10

This does not necessarily argue, however, against a cost of reproduction in these animals. Reproduction may be costly, in fact so costly that only individuals in good condition breed. A balancing of reproduction against physical condition may well result in nonreproduction when an individual's condition is so poor that reproduction might result in mortality. Ten years' data for hoary marmots reveal a tendency for reproducing females to suffer a higher overwinter mortality than nonreproducing females: of 191 observed instances of females aged three years and over entering hibernation, 22 were followed by the female's death that winter, whereas 169 were followed by her survival to emerge the next spring. Of the animals that died, 15 (68%) had reproduced the previous summer, whereas of those that survived, 39 (23%) had reproduced the previous summer.[4] If larger, healthier, more robust individuals are more likely to breed, then a somewhat higher survivorship would actually be expected, not because of their reproducing but, more likely, in spite of it. In this respect, then, individual differences in reproductive capacities may be of special, and generally unappreciated, importance.

In addition, inclement weather may take a special toll of breeding females as well as of their young. A mid-August snowstorm in 1976 deposited about 10 cm of snow in the Washington Cascades, after which adult males, yearlings, and nonreproductive females apparently began hibernation. Young of the year and reproductive females remained above ground for at least several weeks longer, and mortality that winter was 78% (seven of nine) among known reproducing females as opposed to 17% (two of 12) among known nonreproducing females. It seems likely that the onset of winter influences the mortality of yellow-bellied marmots as well since, as we have seen, reproductive females begin gaining weight later than do nonbreeders (Downhower and Armitage, 1971).

Another possible factor in marmot reproduction, not intensively studied among the Nearctic species, is resorption of embryos. In the long-tailed marmot, for example, embryo resorption in some years has been reported to reach 49% of all those produced. High resorption rates of this sort are particularly likely in years with an unusually cold spring (Davidov et al., 1978). More generally, the costs of reproduction among marmots may become especially acute as a result of the vagaries of weather, events that are unpredictable from year to year but which may occur once or more during the lifetime of an average individual.

In addition to its effect on survival *per se*, a more important measure of the reproductive cost of breeding should be its effect on total life expec-

[4] binomial test, P < .05

tancy, if the effects of breeding episodes tend to accumulate over an individual's lifetime. Moreover, the impact of breeding at a given time on fecundity, growth rate, and social status at the next breeding opportunity may well be important. And finally, an individual's reproduction may have a detrimental effect on the success of its previous offspring. Because dispersal among hoary and Olympic marmots seems to be influenced by the emergence of young of the year in spring, a female's reproduction may decrease her fitness via offspring already produced. This should be especially true for species such as *M. flaviventris*, in which female offspring tend to assume the home range of their mothers (Armitage, 1984). In such cases, a failure to reproduce could actually reflect parental investment in young of the previous year (the current yearlings). If this analysis is correct, the alternate-year nonbreeding of *M. caligata* and *olympus* may in itself reflect a type of parental investment, just as their alternate-year breeding represents more typical reproductive effort.

Orthodox evolutionary theory suggests that among long-lived iteroparous (repeatedly breeding) species such as marmots, reproductive effort should increase with increasing age, since as residual reproductive value declines there is increasing advantage to maximizing short-term gain. To some extent this is confirmed, in that a smaller proportion of woodchucks breed as yearlings, yellow-bellied marmots as two-year-olds, and hoary and Olympic marmots as three-year-olds, than is the case for older individuals of each species. Moreover, in their initial breeding efforts young marmots tend to produce smaller litters than do older adults: in *M. monax*, yearlings average 3.13 young per litter, adult females, 3.97 young per litter (Snyder and Christian, 1960); in *M. caligata*, three-year-old females averaged 2.3 young per litter (SD = 0.5), older animals, 4.2 young per litter (SD = 0.7).[5] Beyond the initial breeding year, however, there is no trend in *M. caligata* for increased litter size with maternal age (Fig 12.2).

The proportion of breeding effort devoted to sons or daughters can be considered a special aspect of reproductive strategy. Evolutionary theory suggests that sex ratios should approximate $1:1$, if males and females are equally costly to produce, because the rarer sex would experience a mating advantage that would tend to equalize the ratio (Fisher, 1930). However, although overall sex ratios are often $1:1$, nutritional, social, and ecological factors often produce some deviation within a population, and various theories have attempted to explain these cases (see review in

[5] Mann-Whitney U tests in each case, $P < .05$; n = 26 three-year-olds and 21 older animals

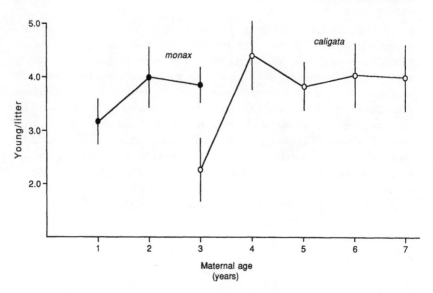

Fig. 12.2. Correlation of maternal age and litter size for *M. monax* and *M. caligata*. Data for *M. monax* from Snyder and Christian (1960); data for *M. caligata* based on a minimum of five females in each age class.

Williams, 1979). Notably for the case of marmots, Trivers and Willard (1973) suggested that in polygynous species, well-nourished and/or dominant females (whose offspring are likely to be unusually successful) should produce an excess of sons, and poorly nourished and/or subordinate females should produce an excess of daughters. In addition, the theory of "local resource competition" (Clark, 1978) suggests that females that are facing potential competition from their daughters should bias their production toward sons.

Data from yellow-bellied marmots do not support either of these theories. Although the sex ratio does not differ significantly from 1 (as for all marmot species so far as known) and the species is polygynous, the following holds for *M. flaviventris*: neither litter size nor stress—measured by eosinophil concentration and mirror-image stimulation—is correlated with biased sex ratios. Over their lifetime, subordinate females produce an excess of males (contrary to the Trivers-Willard theory), and young females that are members of matrilineal social groups produce an excess of females, contrary to the local resource competition theory (Armitage, 1987). This suggests a simpler, more straightforward interpretation: females tend to favor the sex that has the higher probability of reproductive success and that will provide them with a higher number of

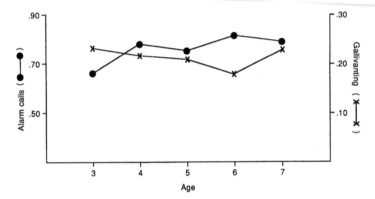

Fig. 12.3. Correlation of alarm calling and gallivanting with age of adult male hoary marmots (*M. caligata*). Data are presented per individual per observation-hour, and are based on a minimum of four different individuals for each age class and a minimum of 81 observation-hours for each age class.

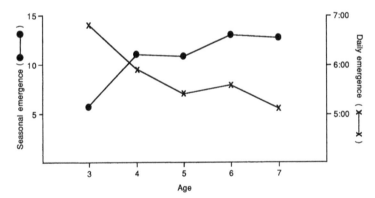

Fig. 12.4. Correlation of daily and seasonal emergence with age of adult male hoary marmots (*M. caligata*). Seasonal emergence is presented as the number of days before the mean emergence of the adult in each male's colony; daily emergence times apply to weeks 1–5 after seasonal emergence. Based on a minimum of four individuals in each age category and, for daily emergences, a minimum of four different times for each individual.

grandchildren. This leads to a bias for males among subordinate females who are unlikely to establish a matriline and who must therefore rely on possible success by a disperser, and a bias for females among matriline members who are likely to recruit their daughters successfully (Armitage, 1987).

We have seen that among females there is no apparent correlation between reproductive effort and age. A similar result obtains for adult males, as measured by frequency of alarm calling or gallivanting excursions (Fig. 12.3). On the other hand, there are positive correlations between the age of adult males and the time of daily emergence during the breeding season and between the age of adult males and the date of their seasonal emergence: older males emerge earlier, both daily and seasonally, than younger ones (Fig. 12.4).[6] This may be a form of reproductive effort in that, presumably, such early-morning and early-season vigilance and activity is metabolically costly; also, early emergers may be more at risk to predators (Fig. 12.5). There is, however, an important confounding factor, namely, the occasional presence of satellite males. When a satellite is present, colony males emerge from their burrows and begin patrolling their colonies earlier in the morning and earlier in the season than do colony males without satellites, and, as we have seen, older colony males are more likely to have a satellite.

When the data relating male age and daily and seasonal emergence are reanalyzed to consider only males that do not have satellites, the age effect disappears. We can conclude, therefore, that older males are *capable* of greater reproductive effort when such effort is appropriate, but there is no reason to suspect that younger males would not behave similarly if they encountered a similar provocation.

Patterns of Interspecific Variation in Reproductive Effort

The different marmot species show different suites of reproductive effort, patterns that reveal a consistent trend relating reproductive effort and behavioral biology to other life-history features. Woodchucks mature as yearlings and reproduce annually thereafter. Yellow-bellied marmots mature as two-year-olds and reproduce irregularly thereafter. Hoary and Olympic marmots become sexually mature as three-year-olds and reproduce biennially thereafter (Fig. 12.6); theirs is the lowest reproductive rate reported for the usually prolific rodents. And since females occasionally skip additional years, the actual reproductive rates of *M. caligata* and *olympus* are even lower.

Reproductive patterns of other marmot species have not yet been carefully documented, but from at least one preliminary study, it appears that *M. vancouverensis* reproduces biennially as well (Heard, 1977). *M. broweri*, like *M. olympus* and *caligata*, do not become sexually mature until their

---

[6]Spearman rank tests, P < .05

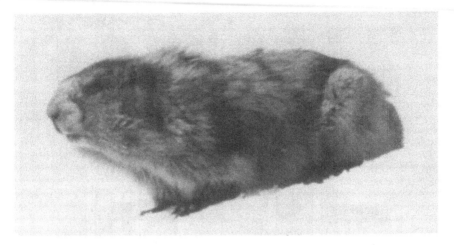

Fig. 12.5. Adult male Olympic marmot (*M. olympus*), one day after seasonal emergence in early May. The animal is very conspicuous against the snow. (Photo by D. P. Barash)

fourth summer, when they are three-year-olds (Loibl, 1983). The reproductive pattern is not known for any of the Eurasian species, although in *M. bobak* the percentage of breeding females varies from 52 to 89% in different years (Shubin, Abelentsev, and Semikhatova, 1978). *M. bobak* (Ismaghilov, 1956), *caudata* (Kizilov and Berendaev, 1978), and *menzbieri* (Maschkin, 1982) are reported to breed initially as three-year-olds, the latter occasionally as two-year-olds, and at least some adults of *M. menzbieri* continue to breed in their thirteenth year. Litter size in *M. camtschatica* has been reported to be as large as 11, although five is the median (Kapitonov, 1960); it is not clear, unfortunately, whether this species breeds biennially, but because of the short growing season and permafrost of its environment, this seems likely. I found both young of the year and associated yearlings in several colonies of *M. marmota* (Barash, 1976a), suggesting annual reproduction in this species, and it is interesting to note that the litter sizes were unusually small (mean = 2.4), suggesting that in the Alpine marmot, annual reproduction may be combined with smaller reproductive effort per year.

Conventional wisdom among Alpine people has it that *M. marmota* is a recent immigrant to the high mountains, having been driven into the talus and meadows by substantial hunting pressure in the lowlands, paralleling the experience of other medium to large European mammals (Couturier, 1964). If so, then annual reproduction in this species could

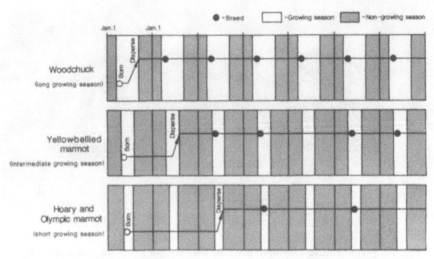

Fig. 12.6. Patterning of maturation, dispersal, breeding, and growing season among some North American marmots.

simply represent the continuation of a pattern adapted to a different environment than the one currently occupied.

In central New York State, W. J. Hamilton (1934) found that 43 of 44 adult female woodchucks reproduced, as opposed to 11 of 43 yearlings. Among adult males, 34 of 38 had enlarged and descended testes, whereas only four of 45 yearling males showed similar signs of sexual maturity, and even then not until May, by which time the breeding season was effectively over. All three-year-old male hoary and Olympic marmots have enlarged and descended testes; no two-year-olds or younger males have them.

Reproductive rate is the number of young born per adult female in the population. The reproductive rate among woodchucks was determined in three different years to be 2.59, 2.57, and 3.24 (Snyder and Christian, 1960). By contrast, yellow-bellied marmots at 2,900 meters elevation in Colorado average a reproductive rate of 1.99 (Armitage and Downhower, 1974), and hoary marmots in Alaska average 0.64 (data from Holmes, 1984b) and in Washington's Cascade Mountains average 1.1. Thus, there is a general trend for reproductive rates to be highest among smaller-bodied species that occupy low-elevation environments with a long growing season. Reproductive rates are reduced among larger-bodied, higher-elevation inhabitants of environments with a shorter growing season.

Intraspecific variation in reproductive patterns has been documented

most thoroughly for *M. flaviventris*; such variation may well be most developed in this species, which occupies environments ranging from low-
elevation arid (in the cold deserts of eastern Oregon and Washington) to
high-elevation alpine (in the Sierra Nevada, southern Cascades and central Rocky Mountains). Within this species, litter size tends to be larger at
lower elevations. Thus, litter size at 2,117 meters elevation in Yellowstone National Park averages 5.26 (Armitage and Downhower, 1974),
and at 1,980 meters in the Sierra Nevada it averages 5.23 (Nee, 1969),
whereas the same species at 2,900 meters in the Colorado Rockies averages 4.15 for colonial females and 4.46 for peripheral females (Armitage
and Downhower, 1974). Among a high-elevation population at 3,400 meters in Colorado, no adult females weaned young in consecutive years, and
litter size averaged 4.07 (Johns and Armitage, 1979).

It is interesting that this intraspecies trend does not hold between species: woodchucks, occupying the lowest elevations with the longest
growing season, average 3.9 young per litter in central Pennsylvania
(Snyder and Christian, 1960), whereas the high-elevation Olympic marmots average 4.0 young per litter (Barash, 1973c) and the equally high-
elevation hoary marmots also average 3.9. Nonetheless, environmental
conditions clearly influence the patterns of marmot reproductive effort,
such that woodchucks are adapted to reproduce early and often, yellow-
bellied marmots less early and less often, and hoary and Olympic marmots the latest and least often. Because of the apparently adaptive adjustment of age at first reproduction as well as the breeding regime in each
species, however, it appears that the marginal cost of varying litter size
is greater than the benefit. (*M. marmota*, as discussed earlier, may be
an exception.)

When a species such as *M. flaviventris*, which is adapted to an intermediate reproductive regime, experiences variations in environmental
conditions, modifications in litter size are apparently more easily achieved
than variations in age at first breeding or the annual patterning of reproduction itself. It is therefore interesting that whereas peripheral female
yellow-bellied marmots tend to have larger litters than do colonial females, they also skip reproductive years more often (Armitage and
Downhower, 1974). Their reproductive performance may exemplify the
pattern of reproductive suppression of subordinates described above for
hoary marmots, although they have less opportunity for direct behav-
iorally mediated suppression. Such individuals may also be affected more
by snow conditions and hence not be physiologically able to reproduce as
often (K. B. Armitage, personal communication). In this case, the effect
would still be an ultimate consequence of interindividual competition, if

competitive factors are responsible for a female's peripheral status, even though the proximate cause of the reproductive pattern may involve non-social factors.

The Ecology of Paternal Behavior in Hoary Marmots

Reproductive effort involves more than simply pumping nutrients into one's offspring. Parental behavior in general constitutes reproductive effort, although such efforts are generally difficult to quantify or occasionally even to recognize. In some cases, the expenditure of reproductive effort via parental behavior occurs at the cost of the parent's ability to invest in subsequent offspring (parental investment, *sensu* Trivers, 1972). In others, parental behavior may interfere with the parent's ability to achieve additional matings. In practice, most field studies of parental investment take a less rigid perspective and are concerned with any behavior that benefits offspring, simply because subsequent costs can only rarely be clearly identified or quantified. In any event, whereas decrements in ability to invest in future offspring seem especially likely to impinge on females' optimal reproductive strategies, decrements in subsequent mating opportunities seem especially relevant to males.

I observed the paternal behavior of adult male hoary marmots, comparing the behavior of adult males at three isolated colonies (two monogamous, one bigamous) in Mount Rainier National Park, Washington, with that of three adult males at a colony town consisting of three adult males, five adult females, and their associated offspring (Barash, 1975b). In all cases, each male was associated spatially with a different litter of young. Thus, the primary sleeping burrow of each adult male was within 50 meters of the sleeping burrow of a litter of young and farther than 100 meters from the burrow of any other litter. No interactions were observed between the young of different families.

Not surprisingly, there were significant differences between the behavior of adult males and of adult females toward their offspring; of greater interest, there were significant differences between the behavior of adult males at the isolated colonies and those at the spatially continuous one. These differences suggest an adaptive pattern to the ecology of paternal reproductive effort in this species. Adult males at the colony town averaged significantly less time within 3 meters of a young marmot than did nonmaternal adult females, who in turn spent less time near young animals than did the mothers (Fig. 12.7).[7] By contrast, the adult males at the

[7] t tests, $P < .05$

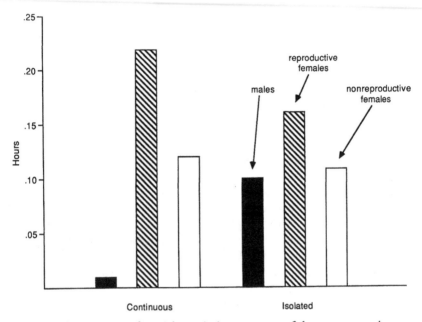

Fig. 12.7. Comparison of spatial proximity to young of the year at continuous and isolated *M. caligata* colonies. Data are presented as time within 3 meters of a young of the year, per observation-hour; based on 127 observation-hours at the continuous site and 137 observation-hours at the isolated sites.

three isolated colonies spent on average significantly more time near their young than did the colony-town males, although still less time than the mothers[8] at isolated colonies. There were no significant differences in mother/young proximity at the two types of sites.

These findings suggest that adult males at continuous colonies exert less reproductive effort—in the form of paternal solicitude and association—than do their counterparts at isolated colonies. In addition, data on greetings reveal a similar pattern: adult males at the three isolated colonies devoted an average of 31% of their greetings to young of the year, whereas young accounted for only 8% of the greetings by colony-town males.[9] This difference in paternal behavior could be due to behavioral differences between the young at the two colony situations, so I obtained data on greeting frequencies by the maternal females. Just as with the proximity data presented above, regardless of the colony situation, no

[8] t tests, P < .05
[9] binomial test, P < .05

difference in the behavior of maternal females toward their young is apparent: 39% of mothers' greetings at isolated colonies were with their young, vs. 34% at the colony town.

These data on greeting and proximity for maternal females in both situations therefore serve as a control of sorts, suggesting that whereas maternal behavior at isolated and continuous colonies is similar, paternal behavior is not. This greater intensity of paternal involvement at the isolated colonies correlates with males' generally higher level of social interaction with individuals other than young at the colony town. Thus, the colony-town males engaged in a total of 0.86 greetings (SD = 0.24) and 0.21 chases (SD = 0.10) per observation-hour, whereas the isolated colony males averaged 0.49 (SD = 0.27) greetings and 0.07 (SD = 0.05) chases.[10] The lower degree of paternal involvement at the colony town therefore correlates with a high level of social interaction with other adults, whereas the higher degree of paternal involvement at the isolated colonies correlates with a low level of social interaction.

Each of the colony-town males was spatially associated with a female and her litter, although they did not share the same burrow. These males averaged 0.56 greetings (SD = 0.38) and 0.04 chases (SD = 0.03) per hour when greetings and chases involving animals outside this nuclear group were excluded from the analysis. Thus modified, the data on the colony-town males do not differ significantly from those on the isolated-colony males, supporting the hypothesis that the reduced social activity of the isolated animals may have been due to their colony situation rather than to a reduced social tendency *per se*. The increased paternal involvement of isolated-colony males may well be a result of their generally reduced social interactions with other individuals; similarly, the low paternal involvement of colony-town males may result from the demands of their intense social activities with other adults (see Chapter 8).

Male mammals often cannot make a substantial direct contribution to the eventual reproductive success of their offspring; specifically paternal care is therefore rare among mammals. In contrast to female fitness, male fitness is often maximized by a constellation of behaviors directed toward both inter- and intraspecific defense. Assuming a male's fitness is reduced if he permits a strange male to inseminate "his" female and increased if he copulates with a neighboring female, it is not surprising that males living around other adults—such as those at a colony town—are inclined toward a high degree of social involvement. Under these conditions, di-

[10] t tests, P < .05

rect paternal investment is of much lower priority. Conversely, males living in relative isolation are "free" to expend additional reproductive effort on their offspring directly, both because the need for watchfulness and the recompense for gallivanting are vastly reduced.

Because estrus occurs during a restricted period shortly following emergence in the spring, it may seem surprising that even the colony-town males do not revert to doting fatherhood each summer, once their reproductive interests are secured. As we have seen, a significant seasonal decline in social interactions among adults is characteristic of marmots of all species, although it may well be advantageous for adult males to remain somewhat alert for the intrusions of dispersing two-year-olds, more likely at continuous than at isolated colonies.

The preceding analysis assumes that direct paternal effort is somehow advantageous to the young and hence that it contributes in a real though small way to the inclusive fitness of the male. As one possibility, male involvement may result in greater social integration of the young. Among yellow-bellied marmots, for example, peripheral residents disperse as young of the year and are not found in succeeding years; their survivorship is low, possibly zero. At colonial sites, by contrast, the young always overwinter and do not disperse until the next year, as yearlings (Svendsen, 1974). Because animals in peripheral situations often lack a resident male, their earlier and presumably "less fit" dispersal may be a partial consequence of male noninvolvement. Thus, enhanced paternal involvement at isolated hoary marmot colonies may be a partial adaptation to prevent the liabilities of male deprivation suggested for *M. flaviventris*.

Alternatively (or additionally), adult males may benefit the young by helping them avoid predators. Because of their small size, young marmots are more susceptible to predation by foxes and *Buteo* hawks than are larger animals. In addition, they are often preoccupied with vigorous play-fighting and are slow to enter their burrows at the approach of danger. On three different occasions, I witnessed young marmots being physically induced by an adult (their mother) to enter their burrow as I approached. Adult males certainly appear equally capable of such behavior, and their close spatial association with the infants at the solitary colonies would clearly enhance their prospects for doing so. Ten times I saw adult males enter their burrows in potentially threatening circumstances, whereupon the infants immediately did the same, although there was no indication that the male acted directly upon the infants to produce this behavior.

It is also possible that there are more "avoider" personalities (Svendsen, 1974) among male inhabitants of isolated colonies than at colony towns— that is, individuals that avoid other adults but do not avoid young. Alternatively, the pattern of paternal effort—or more commonly, noneffort— described here may be a simple function of the physical arrangement of individuals, isolated-colony males interacting with young animals simply by default, because few others are around, and not because of any particular fitness benefit they accrue. It seems clear, in any event, that reproductive effort among adult male marmots occurs largely via interactions with other adults, in which the males seek to maintain and inseminate as many adult females as possible and that patterns of paternal investment, although consistent with an evolutionary interpretation, are not especially robust.

Among the colonial species, physical proximity between adults and their offspring provides the opportunity for prolonged parental investment and reproductive effort by both sexes, even if the investment and effort is largely passive, i.e., varying degrees of tolerance, if not solicitude. In a sense, colonial social systems can be viewed as the temporal and spatial manifestations of reproductive effort extended beyond the mammalian physiologic necessities of gestation and weaning. The simple toleration of weaned young, yearlings, and two-year-olds among hoary and Olympic marmots bespeaks a degree of prolonged investment. Among yellow-bellied marmots, such investment extends typically to young and yearlings (more often yearling females than males) but not to two-year-olds. And finally, among woodchucks, only the young are recipients of reproductive effort, and then only for a brief time.

The expenditure of parental investment by male hoary marmots is problematic; as little as they provide, however, male woodchucks almost certainly provide even less. Among colonial species, parents live close to their offspring. Adult male woodchucks, by contrast, travel some distance from their inseminated females, reducing any opportunity for them to expend reproductive effort on their young. The early dispersal of *M. monax* similarly diminishes the woodchuck mothers' opportunity for investment in their weaned offspring. The rarity of alarm calls, in both sexes, may also reflect this spatial/behavioral factor, since even for woodchuck mothers, the opportunity for reproductive effort via alarm calling is reduced. Although total lifetime reproductive effort may if anything be greater in the more intolerant, rapidly maturing, and rapidly breeding species, parental investment in any given young of the year or reproductive effort per litter is higher in proportion as the species is colonial, late maturing, and late dispersing.

Foraging as Reproductive Effort in Hoary Marmots

An evolutionary perspective seems most striking when it elucidates complex social patterns. The excitement generated by sociobiology was due at least partly to the fact that behavior such as altruism, care of young, and choice of mates seems unlikely, at first blush, to be acted upon directly by natural selection. Recognizing the legitimacy of applying evolutionary concepts to social behavior, however, ought not to obscure their likely relevance to seemingly "nonsocial" maintenance behaviors as well. Evolution by natural selection should be no less valid, if less eye-catching, in the latter case than in the former.

Foraging by a parent marmot, for example, is reproductive effort insofar as it contributes to reproduction. Moreover, it constitutes parental investment as well, insofar as it increases the chances of offspring survival at the potential cost of the parent's ability to invest in additional offspring (Trivers, 1972). Foraging by a reproductive female marmot can constitute parental investment when it takes place in a manner that reduces the nutrient input to the parent, increases the parent's risk of predation, or in some other manner reduces the parent's ability to invest in future offspring. In return for these costs imposed on the parent, parental foraging increases the nutrient input for the offspring, reduces the offspring's risk of predation, or in some other manner increases the chances of the offspring's survival and hence reproduction. Accordingly, parental marmots can be expected to show signs of parental investment not evident among comparable adult nonparental marmots. Without the compensating benefits derived via increased success of one's offspring or other relatives, selection should act against foraging (or indeed, doing anything) in a manner that is in any way disadvantageous for the individual concerned. Individuals should avoid behaving in a way that reduces their residual reproductive value, unless the benefits gained via their current reproductive value are sufficiently great to offset this cost.

Reproductive female hoary marmots in May and June have a higher reproductive value than do nonreproductive females at the same time. Because of the metabolic demands of pregnancy and lactation, reproductive females (rfs) at this time should have greater foraging needs than should nonreproductive females (nrfs). However, herbivores are generally limited by the rate at which they can process food in their gut rather than the rate at which they capture nutrients, since they feed on immobile foods of low caloric density and must balance micronutrient intake as well (Westoby, 1974). The typical foraging pattern for marmots is to feed in a series of bouts throughout the day, separated by within-burrow periods during which digestion is presumably occurring and individuals are relatively safe from predators. The following predictions regarding the foraging behavior of rfs and nrfs were made and tested over a four-year period, with data gathered at four different colonies of hoary marmots (Barash, 1980a).

Prediction 1. In June, when rfs are in late pregnancy and early lactation, rfs will run more risks than nrfs to meet their presumed greater foraging needs. In other words, rfs should increase their reproductive effort, at some possible cost to their residual reproductive value, when their present reproductive value is high relative to the present reproductive value of nrfs.

Test. I obtained a measure of foraging activity by censusing each adult female every 5 minutes and recording the number of seconds she spent foraging during the ensuing 1 minute. By dividing the total number of seconds foraging by the total observation time, I obtained a foraging index—the percentage of time spent foraging by rfs and nrfs. To avoid possible bias in foraging percentage as a function of time of day, observation-hours for rfs and nrfs were balanced during each hour interval.[11] Table 12.1a contrasts rf and nrf foraging activity between 0800 and 2000 hrs under fair skies, partly cloudy skies, or high overcast. Contrary to the prediction, there were no significant differences between these two classes of adult females.

However, when foraging behaviors of rfs and nrfs between 2000 and 2130 hrs are contrasted (by 2130, all marmots are below ground), foraging of rfs is significantly greater than that of nrfs in June (Table 12.1b).[12] A comparable pattern is revealed when daytime foraging of rfs and nrfs between 0800 and 2000 hrs on days of heavy fog, rain, or snow is compared (Table 12.1c).[13]

Prediction 2. The differences in nutritional needs between rfs and nrfs should be less in August than in June. By August, even parental care is greatly reduced from June and July, since the young are weaned and nutritionally—although not socially—independent. Furthermore, although rfs may be under selection for increased late-season foraging to compensate for the costs of pregnancy and lactation just incurred (Barash, 1976b), there is minimal difference in current reproductive value between rfs and nrfs at this time. (This entire treatment differs greatly from that usually employed in studies of optimal foraging, in that both

[11] chi-square test = 1.04; 11 df
[12] binomial test, $P < .05$
[13] binomial test, $P < .05$

TABLE 12.1

Percentages of Time Spent Foraging by Five Different Reproductive and Nonreproductive Female Hoary Marmots During Four Different Years, in June and August

a. 0800 to 2000 hrs, During Only Fair, Partly Cloudy, or High Overcast Weather; Based on a Minimum of 250 Observation-minutes per Animal per Month

	June	August
Reproductive females	28	34
Nonreproductive females	25	31

b. 2000 to 2130 hrs, During Only Fair, Partly Cloudy, or High Overcast Weather; Based on a Minimum of 50 Observation-minutes per Animal per Month

	June	August
Reproductive females	13	9
Nonreproductive females	4	11

c. 0800 to 2000 hrs, During Fog, Rain, or Snow; Based on a Minimum of 50 Observation-minutes per Animal per Month

	June	August
Reproductive females	9	2
Nonreproductive females	3	3

pursuit and handling time—typically considered important in foraging models—are essentially zero. Furthermore, this approach emphasizes *risk* [Holmes, 1984a] as a significant consideration in optimal foraging.)

Test. Contrary to the prediction, there is no difference between the foraging indices of rfs and nrfs in June versus August (Table 12.1a). But whereas rfs and nrfs differ in foraging from 2000 to 2130 hrs during June, these differences disappear in August (Table 12.1b). Similarly, whereas rfs and nrfs differ significantly in foraging indices during inclement weather in June, these differences also disappear by August (Table 12.1c). Thus, insofar as rf and nrf behavior differs in June, that difference disappears in August.

Prediction 3. Because the fitness of rfs is dependent largely on the success of their young of the year, which have their own nutritional needs to satisfy via foraging, rfs can be predicted to modify their foraging so as to minimize competition with their offspring and, if need be, to accept greater risk in the process, whereas comparable modification of nrf foraging should not occur.

Test. At 15-minute intervals during both June and August, I indicated the position of rfs and nrfs on enlarged photographs of the study sites and measured the distance from each animal to the nearest burrow. Marmots typically retreat to a burrow when confronted by a predator; they fight only as a last resort. Distance to the nearest burrow should therefore be another measure of risk incurred, with risk increasing more or less linearly with the distance. On the other hand, forage availability also seems to increase with greater distance from a burrow, as the depleting effects of foraging by other marmots is felt less strongly (Holmes, 1984a). During June, rfs feed significantly farther from the nearest burrow than do nrfs— 11 meters (SD = 3) versus 5 meters (SD = 1).[14] This is consistent with the expectation that they will run greater risks when experiencing greater need. By August, the differences are once again insignificant—7 meters (SD = 2) versus 8 meters (SD = 2)—consistent with prediction 2 as well.

Prediction 4. Reproductive females should do more alarm calling, spend more time near their burrows, and look up more times per minute while foraging than do nrfs. They should exhibit a suite of behaviors not shown by nrfs that reflect reproductive effort expended on their current offspring.

Test. All these predictions were confirmed (Table 12.2).

The higher proportion of time spent by rfs near their littering burrows suggests that reproductive investment related directly to care of young might interfere with pressures on mother marmots to reduce forage competition with their own offspring. Accordingly, I tested whether rfs actually foraged more than nrfs close to their home burrows by placing wooden stakes at measured distances from both natal burrows and the burrows of nrfs, so that rfs and nrfs could be assigned to one of three concentric rings: less than 5 meters from the burrow, between 5 and 15 meters, and greater than 15 meters. I censused rfs and nrfs every 5 minutes, assigned them to one of the three distances from their burrows, and then obtained a foraging percentage for each case, using the same procedures as described above. Again, the hourly observation regime for rfs and nrfs was similar.[15] The results (Fig. 12.8) show that rfs conduct significantly less of their foraging within 5 meters of their burrows than do nrfs.[16] The opposite is true for the zone 5–15 meters from the burrows,[17] and beyond 15 meters, no differences are apparent. This tendency

[14] t test, P < .05
[15] chi-square = .44; 11 df
[16] binomial test, P < .05
[17] binomial test, P < .05

TABLE 12.2

*Alarm Calling, Burrow Proximity, and Time Spent Looking up
by Female Marmots during July and August (SD)*

Status of female	Alarm calling	Burrow proximity (%)	Looking up
Reproductive	0.80 (.3)	45	21 (5)
Nonreproductive	0.33 (.2)	21	11 (3)
	$P < .05$	$P < .05$	$P < .05$
	t test	binomial test	t test

NOTE: Alarm calling = per female per hour. Burrow proximity = % of 5-minute censuses within 5 meters of the nursing burrow. Looking up = seconds per observation-minute per animal.

of rfs to forage less near their burrows than nrfs do near theirs is especially striking in view of the greater proportion of nonforaging time rfs normally spend near their burrows. Thus, whereas nrfs averaged 22% of their census time during July within 5 meters of their burrows, rfs averaged 43%.[18]

It seems likely, of course, that both rfs and nrfs are under substantial pressure to forage effectively, since they are limited to an above-ground vegetative growing season of only about 3 months, during which they must accumulate sufficient nutrients to last through hibernation. Both rfs and nrfs therefore forage at a high level during most days, probably at the maximum that permits effective nutrient absorption in the gut; hence the lack of difference between rfs and nrfs during daytime and in good weather. During June, however, when their current reproductive value is higher than that of nrfs, rfs accept higher costs in order to obtain the needed nutrients. This increased foraging risk can therefore be considered a form of reproductive effort, with costs incurred via an increased risk of predation as well as a possible increased risk of hypothermia. Both of these risks are presumably more significant after 2000 hrs and during inclement weather.

The reduced near-burrow foraging of rfs can be interpreted as avoidance of forage competition and hence a form of parental investment, as rfs must travel farther from their burrows, expending time and energy in the process, while also being presumably at greater risk from predators. Preoccupation with their young, including greater watchfulness for predators, may be the proximate mechanism reducing the near-burrow foraging of rfs. However, when I reanalyzed the data from Fig. 12.8,

[18] binomial test, $P < .05$

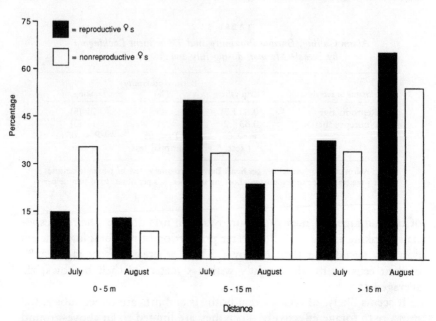

Fig. 12.8. Foraging distances of reproductive and nonreproductive female hoary marmots (*M. caligata*) in July and August, presented as the percentage of foraging times when individuals were within different distances of their home burrows. Based on a minimum of 50 observation-minutes within each distance interval per animal.

comparing rf feeding distances when their young were below ground with those obtained when the young were above ground, there was no significant difference in the results. Thus, rfs reduce their near-burrow foraging relative to nrfs even when their young are not above ground.

Conceivably, reduced near-burrow foraging by rfs could also reflect a somewhat different maternal strategy: allowing plant growth that provides additional cover for the young. In addition, since forage availability increases at increased distance from the occupied burrows (Holmes, 1984a), rfs who concentrate their foraging at greater distances may actually be receiving a greater immediate nutritional return. However, this is consistent with the arguments presented above, since the risks incurred by these rfs are presumably greater as well, just as their likely need for nutritional return is also greater than for nrfs. Along with reducing forage competition with their own young, this could further predispose rfs to forage at greater distances than nrfs. In addition, the difference between forage quality near and far from burrows is actually greater in August than in July, by which time reproductive females are no longer avoiding near-burrow foraging and, of course, their young are wandering widely. Foraging differences between rfs and nrfs have disappeared by August, when the wider foraging radius of their young has presumably reduced the pressures on rfs to reduce their near-burrow foraging. Moreover, meadow desiccation by August generally seems to result in more selective foraging and hence greater wandering by all animals.

Intensive studies of forage depletion in marmot meadows have not yet been undertaken, but significant differences have been found in the standing crop biomasses of enclosed and open plots in meadows grazed by Olympic marmots (Wood, 1973). Although yellow-bellied marmots only consume a small fraction of the total energy available in their meadows, the effects of forage depletion in the immediate vicinity of a burrow could still be significant, especially when combined with the greater predation risk encountered by all marmots—but especially young of the year—when foraging far from their burrows. Unfortunately, data comparable to the above are unavailable for woodchucks, not only because they are difficult to observe, but also because virtually 100% of all adult females are reproductive each year. As a result, any nrf adult woodchucks should be considered moribund. However, I did compare the foraging behavior of five rf yearling woodchucks with that of 5 nrf yearlings during two weeks before and one week after the emergence of the former's young: there were no differences in the percentage of time spent foraging near (less than 15 meters) or far (greater than 15 meters) from their burrows.

As woodchucks reproduce annually, it may be that they are under less intense selection pressure to invest heavily in any given set of young, since the optimum strategy for such a relatively *r*-selected species is to reproduce often but invest little in any given cohort. On the other hand, perhaps forage depletion is less of an issue for *M. monax*, because its lower-elevation meadows may simply be more resistant to grazing pressure. Also, the fact that only one animal typically forages from a given burrow may mean that the woodchuck depletes its forage supply less than the colonial species deplete theirs. Since the young disperse within a few weeks of their emergence, there is also less advantage to enhancing their forage opportunities in the immediate vicinity of the natal burrow. And finally, yearling woodchucks may be less likely than adults to expend large amounts in parental investment, since they have a higher reproductive value than adults and may therefore be selected to expend less reproductive effort in what is only the first of several potential reproductive attempts.

Summary and Conclusions

1. Employing the ratio of offspring weight to adult female weight and correcting for the proportion of adult females reproducing per year, we obtain a measure of reproductive effort, revealing that among the North American species, woodchucks make the greatest effort, Olympic and hoary marmots the least, with yellow-bellied marmots intermediate.

2. The consequences of reproductive effort are apparent in patterns of weight loss: for males following hibernation and for females in delays in weight gain associated with reproduction.

3. Heavier females are more likely than lighter females to reproduce.

4. First-year breeders tend to have smaller litters than subsequent breeders; otherwise, there is no convincing evidence for increasing reproductive effort with increasing age of marmots.

5. Although litter size does not vary consistently from species to species, in *M. flaviventris* there appears to be an intraspecific trend for higher-elevation populations (enduring a shorter growing season) to have lower reproductive rates.

6. Adult male hoary marmots invest (minimally) in their offspring when alternative opportunities for fitness maximization via mate-guarding or gallivanting are unavailable.

7. Parental investment among marmots varies consistently with degree of coloniality and age of sexual maturation and dispersal, such that the more colonial, late-maturing, and late-dispersing species invest more.

8. Hoary marmots reveal a syndrome of reproductive effort in their foraging behavior, with reproductive females foraging at times when nonreproductive females do not; reproductive females also minimize forage competition with their young; these aspects of reproductive effort are generally no longer identifiable by August. Woodchucks do not appear to expend comparable reproductive effort on their offspring.

CHAPTER 13

Dispersal

E ven sedentary animals typically undergo a dispersal phase, and marmots are no exception. Although some individuals—and in certain cases a majority—may remain for all of their lives in the colony of their birth, others disperse: at some point they leave their natal home and seek their fortunes elsewhere. Dispersal is crucially important, yet inadequately understood. It is important because it occurs among all species of *Marmota*, it almost certainly leads to high levels of mortality, and it directly influences the marmots' demographic, social, and genetic environment. Dispersal is a sieve through which all individuals must pass—if for no other reason, because of the need to "decide" whether or not to disperse at all.

The genetic structure of species and populations is determined in large part by patterns of dispersal, combined with the vagaries of mortality and success in entering existing colonies or establishing new ones. The actual makeup of a marmot colony is determined not only by the patterning of reproductive effort but also by whether or not dispersal occurs. Nondispersing individuals are recruited into the colony population; dispersers, by definition, are lost to the colony, although of course they may return or contribute to other colonies. In addition, the age structure of a colony is a direct function of dispersal, reproductive, and mortality patterns: early dispersal produces a colony skewed toward the older age classes, whereas delayed dispersal results in a colony skewed toward the younger age classes. The social structure of a colony is also a direct function of dispersal, in that sociality itself can be seen as a consequence of delayed dispersal (and, sometimes, vice versa). In the extreme case, if all individuals disperse, asociality may result. And finally, depending on the relative reproductive success of dispersing vs. nondispersing individuals, the actual fitness of each individual is a direct function of whether or not (s)he disperses; among adults, it is similarly a function of when and whether his or her offspring disperse.

Comparative Patterns

Olympic and hoary marmots disperse as two-year-olds. This appears to be the case for Vancouver marmots as well (Heard, 1977) and seems to be the general pattern for the alpine and subalpine montane species. The dispersal ages of the Brower's marmot and of the various high-elevation, late-maturing Asiatic species such as *M. camtschatica, sibirica,* and *himalayana* have not yet been reported. Yellow-bellied marmots at 2,900 meters in the Colorado Rockies disperse as yearlings (Downhower and Armitage, 1981). Male yearlings are significantly more likely to disperse than are females, and most dispersers have left their natal colony by mid-July. By August 15, 73% of 62 male yearling yellow-bellied marmots had dispersed (Armitage, 1974); since male two-year-olds are only rarely found to be resident in their colony of birth, some yearlings evidently disperse in late summer or as two-year-olds immediately after seasonal emergence.

Woodchucks disperse as young of the year (W. J. Hamilton, 1934; Grizzell, 1955; de Vos and Gillespie, 1960). The timing of dispersal among woodchucks is variable, ranging from the day of emergence to several weeks thereafter. In some cases, woodchucks delay dispersing until after their first hibernation, as yearlings. In an Ohio population, 75% of the young of the year dispersed; the remaining 25% did so the following spring. Of these late dispersers, 86% were female (Meier, 1985). Individual females whose offspring delayed dispersal during one year did not necessarily produce offspring that delayed dispersing the following year, suggesting that neither hibernaculum availability nor maternal genotype is responsible for the difference.

Comparing dispersal patterns in *M. monax, flaviventris,* and *olympus,* Meier (1985) suggested that "the amount of time that young are active before dispersal is approximately equal in all three species," so the progressively later dispersing age of the higher-altitude species may be a simple function of longer time spent hibernating per year. However, the great majority of woodchucks disperse after 2 months of above-ground activity; yellow-bellied marmots disperse after 4.5 months, and Olympic and hoary marmots disperse after 7.5 accumulated months.

In the woodchuck population that I observed in New York State, three young of the year dispersed within 1 week of their first appearance in

Fig. 13.1. Two young woodchucks (*M. monax*), five days after seasonal emergence. They dispersed within a week of this photo. (Photo by D. P. Barash)

mid-June, six more dispersed within the following 2 weeks, and ten more during the following 4-week period (Fig. 13.1). By August, all young except one had left their natal burrow. The exception was a female young of the year whose mother was shot by a hunter in mid-July and whose two sisters and one brother had already dispersed. She remained as a resident in her natal burrow.

As with other behavior patterns, dispersal also seems to be more variable in *M. flaviventris* than in any other marmot, perhaps because of the wide variability of habitats this species occupies. Although it is yearlings that disperse in the medium-elevation populations that have been most carefully studied, high- and low-elevation populations reveal consistent differences. At medium elevations (1,700 and 1,980 meters) in the Oregon Cascades, yellow-bellied marmots disperse as yearlings, as they do in Colorado, but at low-elevation xeric sites in Oregon (840 meters), individuals leave their natal colonies as young of the year (Webb, 1981); the same occurs at low-elevation xeric sites in eastern Washington (personal observation). High-elevation (3,400 meters) colonies are characterized by a tendency for dispersal to occur at a later age, if at all. Whereas 73%

of male yearling yellow-bellied marmots dispersed by August 15 at medium-elevation colonies (Armitage, 1974), dispersal and mortality combined accounted for only 38% of yearling males at the high-elevation site. And whereas at medium elevations only 5% of male yearlings and 46% of female yearlings were recruited as nondispersing two-year-olds into their natal colonies, at the high-elevation site 24% of male yearlings and 64% of female yearlings stayed on as two-year-old residents (Johns and Armitage, 1979).

Further evidence of intraspecies dispersal flexibility in *M. flaviventris* is found in the behavior of peripheral individuals. Whereas the offspring of colonial females disperse as yearlings, the offspring of peripherals disperse as young of the year; occasionally, they are abandoned by their mothers (Svendsen, 1974).

Just as dispersal is essentially fixed at 100% by young woodchucks and variable by yearling yellow-bellied marmots, it is notably more variable yet by two-year-old Olympic and hoary marmots. Male Olympic marmots appear as likely to disperse as to remain resident. Of 30 two-year-old males that emerged from hibernation during a three-year period, 16 dispersed and 14 remained resident in the colonies of their birth. By contrast, of 30 two-year-old females that emerged from hibernation during the same period, 10 dispersed and 20 did not. There is accordingly a slight tendency for female two-year-old Olympic marmots to remain in their natal colonies rather than dispersing; however, this trend is not statistically significant, although with a larger sample it might be seen as such.

Among hoary marmots, data covering ten years are available on 119 known emerging two-year-olds: 69 females and 50 males. Of the former, 52 (75%) remained as nondispersing residents, whereas of the latter, only 16 (32%) were similarly recruited into their natal colonies. From these data, male hoary marmots are somewhat more likely to disperse than to remain within their natal colonies,[1] whereas females, by contrast, are significantly more likely to be recruited than to disperse.[2] Dispersal among hoary marmots is clearly sex-biased, with males more likely and females less likely to disperse than would be expected from their proportion in the population alone.[3]

When it comes to sex-related dispersal patterns, the overall situation among hoary marmots is thus similar but not identical to that of medium-elevation yellow-bellied marmots: males are more likely to dis-

[1] binomial test, $P < .10$
[2] binomial test, $P < .05$
[3] chi-square test, $P < .05$

perse than to be recruited. However, the tendency for dispersal by male two-year-old hoary marmots is still somewhat less than for dispersal by male yellow-bellied marmots. After 26 years of studying the latter, for example, K. B. Armitage found only two males that became resident in their natal colonies, and both of these returned after initially dispersing. By contrast, dispersal is male-biased but not invariant in *M. olympus* and *caligata.*

Patterns of dispersal of females of the North American montane species are also similar but not identical: in both *M. caligata* and *olympus,* females are more likely than males to be recruited into their natal colonies, but a higher proportion of available females is recruited in *M. caligata* than in *M. flaviventris.* The dispersal pattern for *M. olympus,* not surprisingly, appears to be similar to that of *M. caligata.* By contrast, in *M. monax* all young disperse, regardless of sex.

Proximate Causes of Dispersal

In view of the importance and universality of dispersal among marmots, it is surprising that we know so little about its actual causes, although increasing data are being gathered as to its immediate antecedents. The following section will present some suggestions as to the factors responsible for marmot dispersal.

Among woodchucks, weaning is correlated with an increase in intolerant responses of the mother toward her offspring, most of which occurs within the natal burrow, before emergence of the young (Barash, 1974b). Dispersal by young of the year occurs shortly after their aboveground appearance. As we have seen, an Ohio population of *M. monax* showed a slight tendency for dispersal of females to be delayed more than in a population occupying similar environmental conditions in Pennsylvania. The Ohio animals also appeared to be somewhat more amicable than their Pennsylvania counterparts; Meier (1985) suggests that this difference may reflect the occasional delay in dispersal in the former population.

At present, comparable studies of mother/young pre-emergence behavior are not available for the other marmot species. It is therefore possible that mother/young intolerance, as it occurs among *monax,* is simply a normal aspect of weaning, found in all marmots, rather than a unique stimulus for dispersal in that species. Moreover, even if those species that disperse later were shown to exhibit similar mother/young intolerance at weaning, after which the young emerge, are weaned, but *do not* disperse, it could still be that the avoidance and social intolerance of young wood-chucks causes them to respond to maternal aggression by dispersing, whereas other species "shrug it off," presumably because such early dispersal would be fitness-reducing for them. Since virtually all surviving animals ultimately leave their natal burrow, we cannot correlate the causes of dispersal among woodchucks by comparing dispersers with nondispersers.

However, field observations suggest a role for maternal intolerance: I observed three litters of recently emerged but not yet dispersed young woodchucks in the vicinity of their natal burrows for a total of 77 hours. There were a total of six bouts of apparent play among the young animals; in no cases did the mother participate. Young animals approached the mother on five occasions: four times the mother responded by raising her tail and threatening; once she responded by walking away. Although I did not observe any violent interactions between mother and young, their exchanges certainly were not affiliative, or even cordial. By contrast, Olympic and hoary marmot mothers are much more solicitous of their newly emerged offspring. It is possible that even with maternal indulgence comparable to that of Olympic or hoary marmots, young woodchucks would nonetheless disperse upon emergence, either because of a direct tendency to do so, or because they are simply intolerant of any but the most peremptory interaction. (It would be fascinating, therefore, to observe the development of young woodchucks in hoary marmot litters, or vice versa, if such interspecies cross-fostering could be accomplished.)

Although it is not possible to compare woodchuck dispersers and nondispersers, we can compare early with late dispersers. Nine young woodchucks were known to disperse within 3 weeks of their emergence, whereas ten more did not disperse until weeks 3–7 after first appearing above ground. If animals delay dispersal until they reach a certain size, one might predict a correlation between weight of offspring and timing of dispersal, such that heavier individuals dispersed earlier. However, the correlation was if anything the reverse: seven of the nine early dispersers whose weights were known averaged 1.20 kg (SD = 0.20) during their first week after emergence, whereas seven of the eight late dispersers averaged 1.38 kg (SD = 0.21) during a comparable period. It should also be recalled that among woodchucks, both adults and some yearlings breed, and since adults emerge from hibernation about 2 weeks earlier than yearlings, their young emerge about 2 weeks earlier as well. Of the 14 young woodchucks considered above, ten were the offspring of adults and four, of yearlings. The former averaged 1.39 kg (SD = 0.13), the latter, 1.16 kg (SD = 0.11).[4] Thus, it appears that adults produce larger offspring that tend to disperse later.

Since adults emerge from hibernation about 2 weeks earlier than yearlings, their young are also weaned about 2 weeks earlier. These larger, earlier emerging young experience a longer latency between weaning and dispersal than do the smaller, later emerging young of yearlings. As a result, offspring of both adults and yearlings disperse at about the same calendar dates. Although early breeding appears to be advantageous for most species, for woodchucks there may well be a particular disadvantage to dispersing too early in the season, when vegetation is not yet maximal and when adult aggressiveness due to breeding competition is still high. If so, then early dispersal should be viewed as a liability rather than an advantage.

Because of their somewhat greater size, the offspring of adults may be able to resist the pressures that might generate premature dispersal in smaller individuals; such dispersal would be especially premature for the offspring of adults, since they are born so early in the season.

If the pattern described here is shown to hold with a larger sample size, remaining questions include the following: (1) Is early dispersal in fact disadvantageous? (2) If so, to whom? That is, what is the relative cost to early and late dispersers and to their mothers? (3) To what extent is dispersal a response of offspring to interactions with the mother, and to what extent do individual mothers differ in the dispersal patterns they initiate? (4) To what extent is dispersal a consequence of the behavior of the offspring, reflecting variable resistance to maternal intolerance, offspring intolerance of sibs and/or mother, or the ontogeny of offspring wanderlust? And this is only a partial list.

The proximate causes of dispersal among yellow-bellied marmots have been under study for many years by Kenneth B. Armitage and his students. No clear relationship exists between the number of yearlings emigrating and the number of young added to a colony during any given year, or between yearling dispersal and the number of adults present in a colony (Armitage and Downhower, 1974). Dispersal in *M. flaviventris*, just like other aspects of their social behavior, is not density-dependent (Armitage, 1975). On the other hand, dispersal is influenced by the behavior of adults toward would-be dispersers.

Adult males are generally more aggressive toward yearlings than are adult females, but this difference is statistically significant only in the case of yearling females. Not surprisingly, adults of both sexes are aggressive

[4] Mann-Whitney U test, P < .10

toward yearling males; it is counterintuitive, however, that adult males are more aggressive toward yearling females than are adult females. In addition to being involved in relatively more agonistic interactions with yearlings, adult males are also involved in relatively more greetings and groomings. And, contrary to expectation, the actual dispersal of yearling males is not statistically related to the levels of aggression they receive— although they are less likely to disperse when rates of greeting and grooming (sometimes termed "amicable" interactions) are high. Adult males initiate more agonistic encounters with yearling males than with yearling females (Downhower and Armitage, 1981).

Unlike woodchucks, early dispersing yearling male yellow-bellied marmots weigh more in June, prior to their dispersal, than do yearlings that delay their dispersal (Downhower and Armitage, 1981). Dispersing radio-tracked yearlings move through heavily forested regions, within which forage and cover are presumably unavailable (Shirer and Downhower, 1968). It therefore seems likely that large size and good physical condition contribute to successful dispersal in these animals.

Of course, the same should apply to *M. monax*, although given their longer growing season and more continuous habitat, woodchucks may be under less intense selection in this respect. In any event, it seems likely that the probability of dispersers finding suitable habitat should be an important part of the calculus determining whether or not dispersal is to occur, and if so, at what age—or alternatively, how much duress should be required. Among yellow-bellied marmots studied in eastern Oregon, colonies from which dispersal occurs without antagonism tend to be those in continuous habitat. When suitable habitat is discontinuous, requiring that dispersers travel through miles of inhospitable terrain (i.e., heavy forests or extensive shrub), would-be dispersers are more "reluctant" to leave and, not surprisingly, must be kicked out (Webb, 1981).

If this is a more widespread pattern, then dispersal among woodchucks—for which suitable habitat is relatively continuous—should be more easily induced than among the montane species, which often occupy discontinuous patches of meadow or talus surrounded by inhospitable terrain. The immediate, local availability of suitable habitat may contribute proximately to the early dispersal of young woodchucks. It is difficult, however, to estimate the accuracy and degree of habitat assessment that can be achieved by a newly emerged young *M. monax*, a yearling *M. flaviventris*, or a two-year-old *M. caligata* or *olympus*. Nonetheless, as the local availability of suitable habitat almost certainly influences the ultimate success of dispersers, it seems reasonable that such availability may influence inclinations to disperse in the first place.

It should be noted that virtually all yearling male yellow-bellied marmots disperse, whereas only about one-half of emerging female yearlings eventually do so. Although the role of aggression from adults in initiating dispersal appears problematic, there is growing evidence that affiliative behavior from adults tends to inhibit dispersal, especially dispersal of yearling females. Furthermore, the presence of adults may contribute to dispersal by younger animals, but in complex ways: removal of adults from one colony, for example, did not prevent dispersal of male yearlings, although it did result in substantial recruitment of females. All six yearling females remained to be two-year-olds at a colony from which adults had been removed; by contrast, during previous years, in the presence of adults, only two of 16 became resident two-year-olds (Brody and Armitage, 1985).[5] It appears, moreover, that it is the presence of adult females rather than of adult males that determines whether yearling females will emigrate: when an adult male but no adult female was present at a yellow-bellied marmot colony, nine of nine yearling females were recruited into the colony as two-year-olds, whereas during 22 years when one or more adult females was present, only seven of 20 yearling females were similarly recruited (K. B. Armitage, personal communication).[6]

The situation for yearling males is less clear: whereas females can generally count on reproducing within their natal colony, males in most cases must emigrate in order to be reproductively successful. This may render their dispersal less susceptible to social influence. It has been suggested that social interaction may determine who disperses, as well as the timing of departure, with asocial, avoider individuals more likely to disperse and social interacters more likely to remain (Bekoff, 1977). Among yellow-bellied marmots, however, individual behavioral phenotypes do not predict who disperses and who remains (Brody and Armitage, 1985).

Harem size appears to be involved, as follows: when harems are larger, adults' aggression is directed more strongly at yearling males, whereas in monogamous colonies aggression is more evenly distributed. Monogamous adult females tend to be more aggressive toward yearling females, whereas polygynous adult females direct their aggression more toward yearling males (Downhower and Armitage, 1981).

For adult females, the marginal cost of one additional female in her colony is presumably greater if the adult in question is initially monogamous than if she is already one of several harem members. This might contribute to the above observations. Aggression by the adult male toward yearling females is more surprising, and as yet unexplained, as

[5] chi-square test, $P < .05$
[6] chi-square test, $P < .05$

males would be predicted to encourage recruitment of additional females into their harems. Inbreeding avoidance might be involved, if the father's aggression helps drive away his daughters. Since, as we shall see, he would likely not mate with them if they are recruited into his own harem, it may be adaptive for him to drive them away, since even if mortality among dispersers is high, some occasional success would contribute to the male's fitness as well. Moreover, transient yellow-bellied marmots are relatively frequent, and are likely to be unrelated to the colony male. Since females' fitness tends to be reduced in larger harems, the presence of already recruited female yearlings would probably make transients less likely to join a harem. Males would therefore derive an additional advantage by driving their own daughters away, thus reducing forage competition and making room for new potential mates, who might be more likely to join a less crowded harem.

On the other hand, the observed aggression by adult males toward yearling females may reflect the early onset of sexual patterns in yearlings, since much of the agonistic behavior involves female yearlings running away when approached by an adult male, behavior similar to that of adult females (K. B. Armitage, personal communication). If so, then avoidance of the adult male by yearling females should not be considered agonistic in the same sense as fighting or chasing.

As we have seen, dispersal by male yearlings is relatively insensitive to adults' aggression toward them, whereas females are more likely to disperse if they experience such aggression. Furthermore, nondispersing yearling females tend to be those that maintain overlapping home ranges with resident adult females, notably their mothers (Armitage, 1975). This important finding introduces a novel genetic, demographic, and social consequence of individual variability in the behavioral profiles, since individual adult females may differ substantially in their aggressiveness as well as the size of the home range they maintain.

Several related hypotheses have been suggested to explain the proximate control of dispersal among yellow-bellied marmots. Among them is a possible correlation between the frequency of pregnant females within a colony and aggressiveness toward yearling females: when snowmelt is late (and hence the growing season is short), fewer females become pregnant and so female/female aggression is reduced, leaving male behavior to determine harem size. As a result, harem size eventually increases. In years with a longer growing season, by contrast, more females become pregnant, and, therefore, female/female competition and aggression increase, a higher proportion of yearlings disperses, and harem size declines (Downhower and Armitage, 1971).

More recently, another version has been proposed, emphasizing the effects of individual variability in adult females' aggressiveness and, therefore, in their tendency to recruit their own offspring. Start with a hypothesized small population of basically sociable animals. Then, as Armitage (1975) suggested, "because social relationships are amicable, recruitment of yearling and/or adult females into the population occurs. As the population increases, social interactions increase. Agonistic behavior is relatively high. Because the older animals are sociable, they remain in the colony. Recruitment decreases. Because of the large number of adults, only a potentially dominant animal is recruited. This animal most likely is a yearling who is unable to dominate the adults, but who does not disperse despite social stress. The combination of numbers plus the recruitment of a highly aggressive animal eliminates any further recruitment. The population declines as older females die and the aggressive animal prevents recruitment. The population may decline to a single adult female. Eventually the aggressive female dies or becomes aged and incapable of excluding recruits. The number of resident female adults increases as sociable animals are recruited. The cycle is repeated."

Recruitment of female yellow-bellied marmots has proven to be consistent with this general picture, modified to include a special role for interactions between mother and daughters, varying particularly with the behavioral profiles of the former (Armitage, 1984). Recruitment is twice as likely to occur when a yearling's mother is present; interestingly, it does not matter whether the yearling's mother is reproductive or nonreproductive during that year, suggesting that the earlier pregnancy-related model of recruitment needs to be either modified or abandoned. Recruiting adult females are those that have been resident longer and produced more litters than nonrecruiters. Also, yearling females tend to restrict their spatial movements, using their mother's home range and thereby buffering themselves against female/female competition from other colony members. In the process, they also appear to increase the chances that they will be recruited. By contrast, when the mother dies, immigrant females are more likely to establish themselves; yearling females, deprived of the spatial buffering otherwise provided by their mother, are then exposed to more female/female competition and aggression and are more likely to disperse.

Finally, successful recruiters tend to be those classified as "social" via mirror-image stimulation. Such individuals share space with their offspring more readily and are less likely to be intimidated by other adult females. By contrast, "asocial" adult females recruit far fewer of their own yearlings into the colony population (Armitage, 1984).

Among Olympic marmots, two-year-olds do the dispersing, and, un-like the situation among yellow-bellied marmots, the behavior is density-dependent. Dispersal occurs by mid-July and primarily as a function of colony population size. When the overwintering mortality of young of the year is high and, hence, a colony's population is low, two-year-olds are less likely to disperse; by contrast, when overwintering mortality is low and many yearlings are therefore present in the spring, dispersal is more likely (Barash, 1973a; see Fig. 11.6). There is no apparent differ-

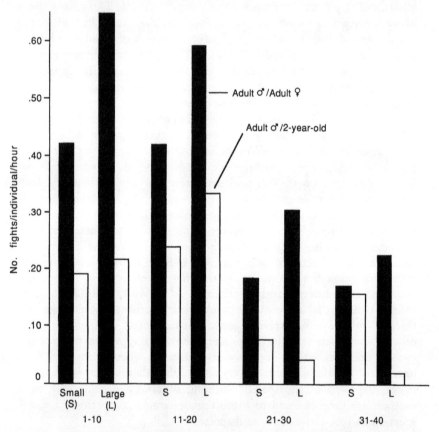

Fig. 13.2. Fights observed between the adult male and the adult females, and between the adult male and two-year-olds at a large and a small *M. olympus* colony. Based on a mean of 43 observation-hours at each colony during each 10-day period.

ence in the density-dependent sensitivity of dispersing male or female two-year-olds.

I conducted intensive daily observations of a small (six animals, no young produced) and of a large (ten animals, two litters of young produced) colony of Olympic marmots. No marmots dispersed from the former, whereas all two-year-olds dispersed beginning at day 25 after emergence at the latter. The two-year-olds fought slightly more with adult males at the large colony than at the small one, although the differences are not significant. Adult male/adult female fights show a similar trend of more fights per individual per hour at the large colony, but here the differences are significant (Fig. 13.2).[7] The large-colony male was involved in significantly more fights than his small-colony counterpart, but his aggression was directed more toward the adult females than toward the two-year-olds. Adult females in both colonies were involved in significantly more fights with the adult male than were the two-year-olds,[8] although it was the two-year-olds that dispersed. The role of adult males' aggression in initiating dispersal of two-year-old Olympic marmots must therefore be considered problematic at best.

Figure 13.3 presents the combined frequency of all fights in which two-year-olds and adult females participated, at both the large and the small colony. Differences between animals at the large and small colonies are apparent but not statistically significant for the two-year-olds, although they are significant for the adult females.[9] These data suggest that two-year-old Olympic marmots engaged in more fights at the large colony than at the small, although the difference was not very great. This "large-colony effect" was more strongly demonstrated by the adult females, who did not emigrate, than by the two-year-olds, who eventually did. A similar difference is found when only the parous females are compared.

Whereas two-year-olds did not engage in more fights at the large colony, large-colony two-year-olds engaged in significantly more greetings than their small-colony counterparts,[10] while the greeting levels of the adult females at the same two colonies were not statistically different (Fig. 13.4). This suggests that for the Olympic marmot increased colony size results in heightened levels of social interaction per individual, as reflected in greeting frequency (and possibly fighting frequency), which corresponds to dispersal of two-year-olds from such colonies.

[7] t test, P < .05
[8] t test, P < .01
[9] t test, P < .01
[10] t test, P < .01

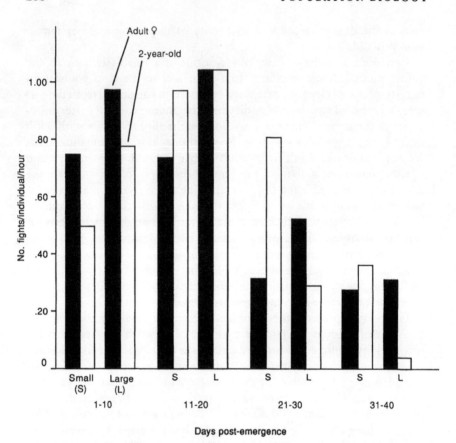

Fig. 13.3. Combined frequencies of all fights experienced by adult females and by two-year-olds at a large and a small *M. olympus* colony. Based on the same observation regime as Fig. 13.2.

The specific role of yearlings in stimulating dispersal is uncertain, although their indirect participation, via their effects on the adults and/or two-year-olds, is probably substantial. Thus, the number of yearlings apparently determines whether a colony will be "large" or "small" the following spring, and it should be emphasized that this number is influenced by the overwinter survival of the previous year's young. The yearlings are generally still underground or just recently emerged at the onset of emigration, and their level of social engagement is generally low for the first week or so after their emergence. It is unclear at present whether the large-colony effect among Olympic marmots is a function of the total number of animals present or more specifically cued by the presence of yearlings themselves. In any event, the overwinter survival of large numbers of yearlings results in a colony's being large, which in turn seems to activate social interactions to which the two-year-olds respond by leaving the home colony.

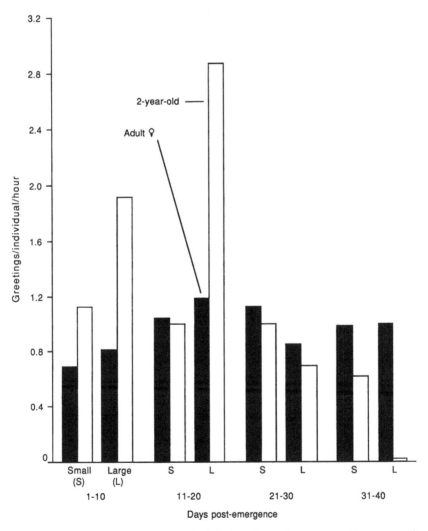

Fig. 13.4. Greeting frequencies of adult females and two-year-olds at a small *M. olympus* colony. Based on the same observation regime as Fig. 13.2.

Dispersal could conceivably be a function of food availability immediately after emergence when the meadows are predominantly snow-covered and greater wandering is necessary to obtain food. Thus, colonies at which overwinter survival and, consequently, dispersal have been low also generally have little snow immediately after spring emergence, because light snow cover raises the winter mortality of hibernating young. However, spring snow depths do not necessarily correlate with the *winter* snow depths that apparently influence mortality. Snow melted quickly from the large colony described above, so forage was not covered at emergence, yet all two-year-olds dispersed. It may be significant that at a north-facing colony with more than 2 meters of snow at spring emergence, Olympic marmots' dispersal was delayed until almost a full month after that at the other sites, by which time forage was more, not less available. This further suggests that snow cover and, hence, food availability are not proximate factors initiating Olympic marmot dispersal. It also supports the correlation among colony size, social pressure, and dispersal described above, since in early spring there was noticeably less social interaction at snowbound colonies and foraging required that animals travel greater distances, more widely spacing them.

Contrary to the usual trend for social interactions to become less frequent as the summer progresses, greeting levels at one north-facing, high-snowpack colony were actually higher in July—when snow had finally melted and the colony residents were foraging in their own meadow—than they had been in June. Fifty-one hours of observation during 4 days in each month revealed peaks of 1.5 greetings per hour in July, as compared to only 0.75 in June.[11] Significantly, dispersal at this colony did not begin until mid-July, by which time greeting levels approximated those of the other colonies, from which two-year-olds had dispersed several weeks earlier.

The proximate motivation for dispersal among two-year-old Olympic marmots may stem from decreased attachment to the colony area as a result of maturation and/or an increased sensitivity to social pressures. Perhaps the sum of these two influences must exceed a threshold for dispersal to occur; this would explain the delay in dispersal at the north-facing colony until the level of social stimulation became sufficient. No adults were known to disperse from any colony. Perhaps by the fourth summer of life, increased attachment to a site and/or decreased sensitivity to social pressure combine to prevent dispersal of mature animals.

[11] binomial test, $P < .05$

Dispersal of Olympic marmots is a gradual affair, with the émigrés becoming progressively more disconnected from the colony and soon sleeping at burrows a hundred meters or so from their home. Upon returning to the main colony area they are often met by a firm rebuff from the residents. These fights, while infrequent and short, are typically vigorous and may contribute to the eventual dispersal. The following selection from my field notes gives the flavor of these interactions: "A two-year-old slept about 100 meters from the main burrow. By 0900 he has been feeding gradually nearer the main colony members; he makes a long run (about 30 meters) toward a burrow but an adult female intercepts him just short of it. She greets him vigorously, followed immediately by an upright fight with much growling. Two-year-old topples over backward and adult chases him about 10 meters; another upright fight, then a chase of about 30 meters. Two-year-old runs off and feeds about 15 meters away."

When the dispersing two-year-old isolates itself from the colony for the night, it misses the "visiting period" of the following morning. As already noted, one possible function of the visiting period and its attendant greetings may be the familiarization of all individuals with the other colony members and the solidification of social relationships. By temporarily isolating itself, presumably in response to the increased level of social interaction at a large colony, the two-year-old may actually precipitate its own dispersal by causing the colony residents to "forget" it—or, at least, cause it to lose its place in the social system. Upon returning to the colony later that day or the next, such a two-year-old is treated much like any strange marmot attempting to enter the colony.

I tested this hypothesis by removing and then reintroducing one two-year-old and two yearlings. As my field notes show, they experienced distinct aggression upon their return: "Two-year-old (tyo) released . . . (after being removed for 24 hours). Adult female is sitting at nearest burrow. Tyo runs right at adult, or perhaps at burrow. Adult approaches and after very brief greeting makes an aggressive lunge with mouth and paw, while growling. Tyo enters burrow rapidly. Adult moves about 1.5 meters away; tyo emerges; adult sees him and approaches; he very actively greets her while her tail goes straight up; tyo quickly reenters burrow."

"Two yearlings, one male and one female, released . . . (after removal for 48 hours). The male enters a distant burrow (and was not seen the rest of the day). The female runs to another burrow. Five minutes later, she emerges and meets an adult female; brief greeting by adult, who growls and chases the yearling at a fast walk, growling and snapping at her heels. This continues for about 30 seconds. Then a long greeting, by adult, with adult growling and yearling raising tail; 15-second walk-chase, adult now growling, then yearling turns and greets adult, very actively. Both animals feed."

To control for the effect of animals being trapped and released, I also released two yearlings and two two-year-olds after holding them in traps for 1–3 hours; there was no notable response by the colony residents, and these animals were immediately reintegrated into the group.

The "walk-chase" that resulted when individuals had been kept for 48 hours was more persistently aggressive than any adult/yearling interaction otherwise observed; presumably the response would have been still more aggressive earlier in the season. In any case, the outcomes suggest that overnight absence from a colony results in a relatively aggressive response by the residents upon return. Thus, if the "large-colony effect" initiates increased wandering by the two-year-olds, the reaction of the residents may cause their break with the home colony to become irrevocable.

Among hoary marmots, dispersal is also moderately density-dependent, but is correlated with behavioral tendencies of the two-year-olds' mothers as well. Although dispersal by hoary marmots is not strongly keyed to the overwinter survival of yearlings, *M. caligata* does reveal a correlation between dispersal and whether resident females bring a litter of young above ground that year: of 119 two-year-olds known to emerge in spring, 51 dispersed. Of these, 44 resided in colonies that produced new litters. Since only 53% of hoary marmot colonies experienced new litters in a typical year, we can safely conclude that two-year-old hoary marmots residing in colonies that produced litters are significantly more likely to disperse than would be expected from chance.[12]

Proximately, this could be due to stimuli emanating directly from the young, although it seems more likely that interaction with the reproductive female(s) is responsible. Although such interactions are generally nonaggressive, they are also so frequent that certain crucial encounters could be of particular importance. Moreover, because such interactions are usually nonaggressive, it is possible that two-year-olds are especially sensitive to aggressive nuances. The frequency of chases is not especially high among hoary marmots, but chases are distributed differently among age and sex classes: adult males and adult females initiate significantly

---

[12] chi-square test, $P < .05$

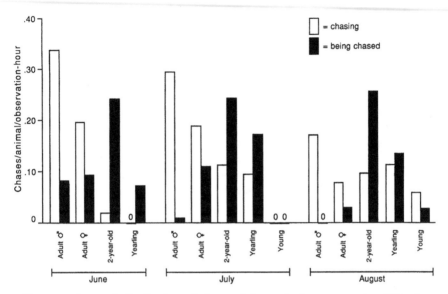

Fig. 13.5. Seasonal distribution of observed chases for hoary marmots (*M. caligata*). Based on a total of 344 observation-hours and a minimum of 4 hours of observation for each hour interval during each month.

more chases than they receive, and two-year-olds and yearlings receive significantly more than they initiate (Fig 13.5).

As reported above for yellow-bellied marmots, male hoary marmots are more likely to disperse than are females: 68% vs. 25%. Dispersal in *M. caligata*, as in *M. flaviventris*, appears to be strongly influenced by the behavioral tendencies of adult females, presumably in concert with the sensitivities of the two-year-olds themselves. Something analogous to the "population-behavior" model proposed by Armitage (1975) appears to hold among hoary marmots as well. I recorded the response of adult females to advances by the adult male, by other females, and by yearlings, categorizing each adult female as "aggressive" or "nonaggressive" depending on whether she stood her ground or retreated when approached. Of 23 females thus categorized, on the basis of 528 observation-hours over four different years, nine were reproductive when tested and 14 were not. Not surprisingly, reproductive females were significantly more aggressive than were nonreproductive females (Fig. 13.6).[13]

[13] Kolmogorov-Smirnoff one-way analysis of variance; binomial tests, $P < .05$ in each case, for all three dyads

It is particularly interesting to note that all of the females that scored higher than 30% in responding to adult males also scored higher than 50% in responding to other adult females and higher than 75% in responding to yearlings. Similarly, all of the females that scored lower than 10% in responding to adult males also scored lower than 50% in responding to adult females and lower than 50% in responding to yearlings. Individual females accordingly maintain a degree of behavioral consistency with regard to their aggressiveness. Of the 34 two-year-olds residing with adult females that were categorized in this manner, nine resided with aggressive and 25 resided with nonaggressive adult females. I examined the number of these two-year-olds that dispersed, comparing the observed number of dispersers with the expected number if dispersal occurred randomly as a function of the number of each category of adult females within the colony. The results showed that two-year-olds residing in a colony with aggressive females are significantly more likely to

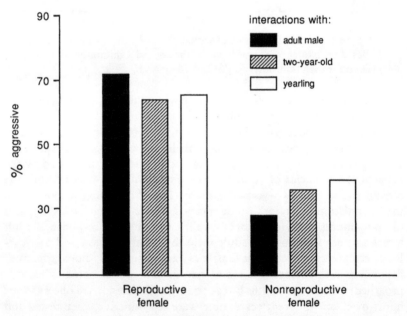

Fig. 13.6. Responses of reproductive and nonreproductive adult female hoary marmots (*M. caligata*) when approached by other individuals. Based on 9 reproductive and 14 nonreproductive females and 528 observation-hours over 4 years.

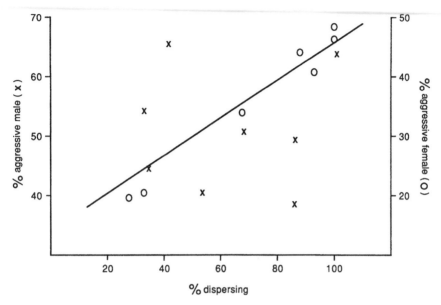

Fig. 13.7. Correlation between dispersal by two-year-old hoary marmots (*M. caligata*) and aggressiveness of their co-resident adult male(s) and adult female(s). Based on 528 observation-hours involving 23 adult females and 11 adult males.

disperse than would be expected based on chance alone, whereas two-year-olds residing with nonaggressive females are significantly less likely to disperse (Fig. 13.7).[14]

By contrast, a similar categorization of colony males (Fig. 13.7) showed no correlation between aggressiveness of the colony male and dispersal, by either male or female two-year-olds. Although adult males vary in their individual behavioral profiles no less than do adult females, it appears that dispersal is more closely cued to the behavioral profiles of the adult females than of the males. This is not surprising, since a three-year-old male, if it remained in its natal colony, would almost certainly be subordinate to its father regardless of the latter's behavioral profile. By contrast, given the existence of female/female competition as described earlier, a three-year-old female's reproductive success may well depend upon the behavioral profile of her co-resident females. This analysis suggests that Armitage's proposed "population-behavioral system" may apply even more closely to hoary than to yellow-bellied marmots, perhaps because of the closer integration of social and colony structure in the for-

[14] chi-square test, P < .05

mer. Such integration would in turn make it likely that the reproductive success of each female hoary marmot is directly contingent on the behavioral profile of every other female. Combined with a slower reproductive rate, this would enhance the intensity of female/female competition and, hence, the impact of female behavioral tendencies on the reproductive success of prospective new colony recruits. In turn, fitness considerations should influence their inclinations to disperse if the probability of their success as recruits is sufficiently low—and, of course, if their opportunity for success as dispersers is sufficiently high.

Olympic and hoary marmots are not rigorously density-dependent when it comes to the initiation of dispersal, but they are clearly density-influenced. In particular, it appears that they are influenced by the type and frequency of interactions they encounter; that is, the "behavioral density" that they experience. This varies as a function of the behavioral profile of the adult female(s) resident in their colony. Two-year-old Olympic and hoary marmots cannot be spatially buffered by the behavior of their mothers as can yellow-bellied marmots, since two-year-olds have complete access to all areas of the colony, regardless of their mothers, and adult females do not maintain distinct home ranges. It is not surprising, therefore, that the proximate stimulus for dispersal in *M. olympus* and *M. caligata* does not come from the pattern of space sharing within the colony, as it does in *M. flaviventris*, but rather from the behavioral tendencies of the adult females in the colony, since two-year-olds are exposed to behavioral interactions with essentially all colony members every day. And these tendencies, in turn, are influenced by pregnancy, and probably by genetic and ontogenetic factors as well.

A cross-species cline may exist in density dependence and other aspects of behavioral sensitivity related to dispersal, with *M. monax* the most independent. At the other extreme, *M. olympus* and *caligata* respond to local population size and/or to the behavioral profiles present in their colonies. Once again, *M. flaviventris* seems to be intermediate, both in the proportion of individuals that actually disperse and in their sensitivity to population and behavioral correlates when they do so.

Inbreeding Avoidance

Inbreeding avoidance is an interesting issue and likely an important one for would-be dispersers, nondispersers, and their parents. Among yellow-bellied marmots, more daughters are recruited into than disperse from the colonies in which their fathers are resident. However, of ten such recruited daughters, nine did not breed, whereas four bred the first year that a new male was present (Armitage, 1984). By contrast, the number breeding and not breeding was not significantly different among two-year-old yearling females whose fathers were *not* present. It therefore appears that the physical and demographic potential for inbreeding exists among yellow-bellied marmots but that inbreeding is avoided, probably because two-year-old females do not undergo estrus if their fathers are present.

Even though woodchucks may breed as yearlings, the potential for inbreeding is presumably lower for *M. monax* than for *M. flaviventris*, because males wander after inseminating a given female, and dispersing young of the year wander as well. Moreover, since habitat suitable for woodchucks is more continuous than that suitable for the montane species, dispersing females probably have more potential living arrangements open to them, allowing them greater opportunity to avoid residing close to their fathers. They are also less likely to end up living nearby if settlement is by chance.

Inbreeding would seem to be an even greater danger for *M. olympus* and *caligata* than for *M. flaviventris*, since the frequency of transients is very low: during three years of study, I never observed a transient (of either sex) gaining entry to an Olympic marmot colony, and during ten years, I witnessed only two transient females entering hoary marmot colonies. As a result, essentially all breeding females of these species reside in their natal colony (excepting those successful dispersers that establish new colonies), compared to a figure of only 48% for breeding female yellow-bellied marmots (Schwartz and Armitage, 1980). In addition, since female Olympic and hoary marmots do not breed before they are three-year-olds, and no more often than every other year thereafter, the cost of inbreeding should be especially high for them, since they have relatively few opportunities to breed and, presumably, can ill afford to produce offspring afflicted with inbreeding depression.

Inbreeding may be less likely among these species because of demographic factors alone. Because both Olympic and hoary marmots do not mature until they are three-year-olds, there is an additional year of potential mortality for their father, increasing the probability that by the time a young marmot is reproductively mature, her father will have died and been replaced. Out of 45 three-year-old hoary marmots whose relationship to the resident adult male was known, only 14 were associated with their fathers. Of these same 45 three-year-olds, 20 reproduced and 25 did not; all 14 three-year-olds whose father was present were among the nonreproducers. Such a distribution would not be expected by chance alone.[15] The following year, the fathers of eight of the original 14 had died or disappeared: of the remaining six still cohabiting with their fathers, only one reproduced. The year after that, two of the previous eight females died, and of the remaining six, only one still cohabited with her father—she did not reproduce, whereas four of the five others (now five-year-olds) reared litters.

It seems likely that for adult female hoary marmots cohabiting with their fathers, the cost of reproducing with him is greater than the benefit of delaying an additional year, because (1) their father is at least three years older than they and hence more likely than they to die during the next year, and (2) they are likely to have a substantial reproductive future, even with biennial breeding. Similarly, just as the cost of reproductive deferral is presumably less than the cost of inbreeding, the cost of dispersal for potentially reproductive females can apparently be greater than the cost of remaining in their natal colony, even if their father is still there as well.

Two-year-old female hoary marmots occasionally disperse, however, and when they do, inbreeding avoidance may be involved. On five different occasions following disappearance of a colony male, I noted the subsequent behavior of his two-year-old daughters. (In three cases I removed the colony male, and twice a colony male died: in all five cases, the colony male was replaced by another male.) There were eight two-year-old males emerging in the spring at these five colonies experiencing male turnover: all dispersed. There were six two-year-old females, and all six emerged into colonies in which new litters were born that year. Under normal circumstances, such two-year-olds can be expected to disperse, whether they are male or female. However, in these cases, five of the six two-year-old females remained in their natal colonies.

The following generalization is suggested, although clearly a larger sample is needed for confirmation: when a two-year-old hoary marmot female emerges into a colony with both a new litter and her father present, she is likely to disperse. When her father is absent, however, dispersal is inhibited, even if a new litter is present. Significantly, replacement of the colony male with another does not have a comparable inhibiting effect on dispersal of two-year-old *males*. Thus, it appears that among hoary marmots, inbreeding avoidance exerts a complex but real influence on female two-year-olds' proximate decision "to disperse or not to disperse." Absent inbreeding effects, dispersal is normally determined by variance in the behavior of the resident adult females.

[15] chi-square, $P < .05$

The Consequences of Dispersal

Dispersing animals may perish, gain entry into an existing colony, or establish a new colony of their own. Dispersers often travel widely and are nearly impossible to track effectively without radio transmitters. It seems likely that mortality among these individuals is very high, although successful individuals may experience high fitness, especially if they succeed in establishing their own breeding colony.

I witnessed three independent examples of new "colonies" established by dispersing two-year-old Olympic marmots; one eventually failed, and the other two apparently succeeded. A single two-year-old male lived alone for an entire summer, whereupon he was joined by a two-year-old female the following spring. The two animals were together that summer but were not seen the next year, presumably having died during hibernation. Two other colonies were formed by dispersing two-year-olds from known colonies. Both new colonies were located within 500 meters of the parent colonies, and both consisted entirely of siblings. One of these colonies produced a litter of young when the inhabitants (two males and two females) were all three-year-olds. In all three cases of dispersers founding colonies, the area chosen was the site of an old, abandoned colony, requiring only minor re-excavation by the new occupants. Old colony sites may be preferred for this reason.

I witnessed the establishment of only one new colony by dispersing hoary marmots: two male two-year-olds (siblings) from one colony and two female two-year-olds (siblings) from another colony established residence at a talus pile during early July. By the following spring, one of the males and both females remained. None bred that year, but the following year one of the four-year-olds produced a litter. The next year, the other original female colonizer (then a five-year-old) also reproduced.

It may seem paradoxical that female hoary marmots apparently avoid breeding with their fathers, while females of both the hoary and Olympic marmots will reproduce with their full-sib brothers. It is possible that these observations are a function of small sample size, or that the siblings in question are actually half sibs. Or, once dispersal has been initiated and a suitable living area has been occupied, the possible liabilities of inbreeding may be less than the cost of deferring breeding, especially since the equal-aged sibs are likely to have comparable lifespans.

Although mortality among Olympic and hoary marmots is in large part density-independent, occurring primarily during the winter, dispersal—as we have seen—is related to population density. Interindividual competition may be severe, especially immediately following emergence when food is scarce and the animals' physiological need may be greatest, not only because of the rigors of hibernation but also because of the reproductive season. The onset of dispersal at this time may also be significant: two-year-olds conceivably fare badly in competition with adults in the spring. So, their likely benefits from dispersal might exceed their benefits from staying home, contending with possible competitors, and perhaps inbreeding.

Among Olympic marmots at least, dispersal appears to be influenced by social factors related to the presence of yearlings, often before they have emerged. Two-year-olds can probably compete successfully with the smaller yearlings, once the latter emerge. It should be emphasized, however, that Olympic marmot colonies, like those of the other montane species, constitute close kinship groups, with two-year-olds and yearlings often being half brothers (having the same father) and/or cousins (their mothers are often sisters). In making the "decision" of whether to disperse or not, two-year-olds should be influenced by selection to minimize competition with their close relatives, possibly by dispersing when the colony population is large. On the other hand, when such competition is low, the balance may be tipped in favor of not dispersing.

Montane marmots inhabit rather small, generally discontinuous meadows, so their numbers may be food-limited when the forage is degraded by an unusually dry summer. Under such conditions, late snowmelt restricts the ability of reproductive females to nourish their young and/or the ability of both the young and their mothers to accumulate sufficient fat to survive their long hibernation.

The fact that dispersal is inhibited when colonies are small implies a lower limit to the optimum colony size, just as the fact that it is frequent when colonies are large implies an upper limit. Possible disadvantages of colony underpopulation include insufficient predator watchfulness and lack of social facilitation, especially of foraging. Activity budgets at a very small Olympic marmot "colony" (one adult male and one adult female) revealed that the inhabitants spent less time being "social" and more time "looking out of burrow" than did residents of larger colonies.[16] Smaller colonies might also suffer greater winter mortality if the body heat of neighboring animals contributes to their survival during hibernation; present data, however, do not show any statistical correlation between overwinter mortality and prehibernation colony size.

Finally, marmots' survival obviously requires substantial biological

[16]binomial tests, $P < .05$

conditioning of their environment. Extensive burrow systems are necessary, not only to provide escape from predators, but also for daily shelter, to provide suitable rearing environments for the young, and to serve as hibernacula. The possible effect of marmots in maintaining suitable habitat for themselves is unknown, but their foraging conceivably destroys developing tree seedlings and effectively prevents succession from meadow to stunted subalpine forest. Larger numbers of individuals would facilitate these benefits and could therefore be adaptive to each participating individual. Excessive numbers, on the other hand, would reduce the fitness of each individual, via resource competition, female/female and male/male competition, and possibly stress effects.

When colonies are large, and in danger of becoming too large, the direct Darwinian benefits of avoiding competition and inbreeding, as well as the indirect inclusive fitness benefit of minimizing competition with close relatives, appear to tip the scales in favor of dispersal. By contrast, when colonies are small, and in danger of becoming too small, dispersal appears to be inhibited, because of the benefit of avoiding the high mortality risk associated with dispersal. In either case, further selection pressure for optimum colony size might also be generated by an inclusive fitness benefit of maintaining a viable colony size for one's relatives.

It should be emphasized that the apparent adjustment of colony size by dispersal in *M. olympus* and/or *caligata* does not require selection operating at the level of the group (= colony), although it does not necessarily preclude this either. It seems most parsimonious to attribute density-dependent dispersal in this species to selection operating at the level of the individual and/or genes.

Dispersal among Olympic and hoary marmots is generally a piecemeal process, the two-year-olds leaving singly or in twos; disengagement occurs gradually and asynchronously over a period of about 5 to 7 days. This gradual emigration may permit precise adjustment of colony size, allowing partial emigration (the leaving of only some of the available two-year-olds) if indicated by the colony size. Since, as we have seen, mortality among colony residents appears to correlate with snowfall—which in the Olympic Mountains is notoriously variable from year to year—mortality is also variable in this species. However, a distinct optimum colony population range seems to exist. It is therefore of interest that the Olympic marmot's behavioral system annually compensates for winter mortality. Genetic adjustments to such short-term alterations would be impossible, whereas a response based on behavior—by definition a relatively plastic, short-term reaction to the immediate environment—is peculiarly well suited to coping with these yearly variations in the population of each colony.

It seems increasingly clear that the behavioral profiles of adult female yellow-bellied and hoary marmots are critically related to dispersal in these species. The dynamic between a yearling and her mother is especially important in the former, whereas in the latter, the crucial interaction appears to be that between two-year-olds and other colony adults, notably the adult females, including but not limited to their mothers. This difference is consistent with the alternate-year breeding system of *M. caligata* as well as the more rigorous partitioning of space by *M. flaviventris*, such that yearlings are significantly more likely to interact with their mothers than with other adult females. Among hoary marmots, by contrast, would-be residents must cope with all other residents. The absence of density dependence in yellow-bellied marmots is rather puzzling, however, just as the extent of density dependence among hoary marmots is difficult to assess.

As to woodchucks, dispersal is virtually complete and appears to be independent of the population's density, its distribution, and the behavioral profiles of the dispersers' mothers (Meier, 1985). The behavioral tendencies of neighboring animals seem less likely to influence the pattern of dispersal for woodchucks than for the montane species, since prior to dispersing, young woodchucks interact almost exclusively with their mothers and each other. Although an aggressive mother woodchuck may well induce early dispersal by her offspring, it is problematic whether aggressive neighbors stimulate emigration or impede it.

Besides influencing colony size and the mortality, survivorship, and reproductive success of individual marmots, dispersal is of great evolutionary importance as the major source of gene exchange between colonies. Among the montane species, patchiness of habitat and persistence of discrete colonies over many generations could promote genetic drift and local differentiation. However, inbreeding avoidance and gene flow between colonies tend to retard the fixation of genetic variants. Exchange rates between colonies are low in *M. flaviventris*, with only 40 successful intercolony transfers noted out of 790 individuals studied over an 18-year period; of these, only 15 seem to have resulted in introgression (Schwartz and Armitage, 1980). It appears that exchange rates for *M. olympus* and *caligata* are even lower, while those for *M. monax* are clearly higher, as gene exchange in that species occurs with every generation.

Summary and Conclusions

1. Dispersal is important to both local demographics and individual fitness, yet it is insufficiently understood.

2. Woodchucks disperse as young of the year; the larger species, inhabiting higher altitudes with shorter growing seasons, disperse progressively later. To some extent, this trend prevails within species as well, at least among yellow-bellied marmots.

3. Dispersal is essentially complete among woodchucks, and progressively less so among the larger species living at higher elevations.

4. Dispersal is strongly sex-biased among yellow-bellied marmots, with males more likely to disperse, females to be recruited into their natal colony; among hoary and Olympic marmots, a similar bias exists, but it is less strong.

5. Among woodchucks, the offspring of adults appear to be somewhat larger than the offspring of yearlings, and the former tend to disperse later.

6. Among yellow-bellied marmots, dispersal by female yearlings is correlated with the availability of suitable habitat and particularly with the nature of the mother/yearling interaction: mothers that share space with their female yearlings are more likely to recruit them.

7. Although dispersal is not density-dependent among yellow-bellied marmots, it is among Olympic marmots, correlating with winter mortality of hibernating young; however, the role of adult aggression in stimulating dispersal by Olympic marmots is problematic, and intolerance of social interactions at colonies containing yearlings may be responsible for some dispersal.

8. Dispersal among two-year-old hoary marmots is significantly correlated with the presence of young that year.

9. Among yellow-bellied marmots, it appears that behavioral tendencies of the mother are especially important, whereas among hoary marmots, the behavioral tendencies of all resident adults—especially the adult females—appear to be relevant.

10. Woodchucks probably experience the lowest probability of incestuous matings, yellow-bellied marmots a higher probability, and hoary and Olympic marmots the highest. However, estrus is apparently inhibited in such cases. Inbreeding avoidance may also influence dispersal, as female hoary marmots, at least, are more likely to disperse when their fathers are present.

11. Dispersal may be influenced by the adaptive advantage of reducing competition, for both the disperser's direct Darwinian advantage and an inclusive fitness benefit.

12. Dispersal is also important as a source of gene exchange between local breeding units; it seems likely that *M. monax* experiences the most gene exchange whereas *M. olympus* and *caligata* are the most viscous, with *M. flaviventris* intermediate.

Basic Patterns and Correlations

In spite of difficulties in determining modal
social systems and mating patterns, for many
species there is considerable uniformity over a
gradient of habitat types in the form of social
organization and the mating system exhibited. It
is the analysis of variations, however, that
will eventually be the key to determining how
social organizations evolve through natural
selection.   John F. Eisenberg (1981)

The human mind seeks order in the natural world, and occasionally the world obliges. In this respect, marmots are reasonably good exemplars of predictable and interpretable trends, coherent and sometimes even gratifying manifestations of natural selection in action. Much of the variance in marmot biology and social behavior seems explicable in sociobiologic terms—that is, as the outcome of individuals behaving in a manner that maximizes the projection of their genes into the future. Furthermore, cross-species comparisons seem to reveal patterns that are consistent with interpretations of this sort. Nonetheless, the natural world is often less orderly than we might like. And here also, marmots are no exception.

CHAPTER 14

The Sociobiology of Marmots: Toward a General Theory

H aving examined marmot social behavior and ecology, I shall now try to present my observations in a broader perspective. Note that my approach is largely inductive; that is, it involves a logical process in which the proposed conclusion involves more information than the data on which it is based. For example, consider the observation "All marmots ever seen (or, more accurately, all those reported in the technical literature) have four legs." This may lead to the supposition "All marmots have four legs." The truth of this statement, however, can only be based on future findings, and cannot be taken for granted. Aside from their descriptive value, such inductive statements as I make in this chapter are useful if they help organize our knowledge and if they lead to more findings, more questions, and yet better theories. They can be useless if they fail to stimulate additional inquiry, and downright hurtful if they are wrong and/or misleading, or if they encourage people to explore blind alleys. Hence, I offer this concluding chapter with speculative optimism, a desire to spur further work, and appropriate caveats, not as a final peroration.

Basic Correlations

Because much of this book has been concerned with identifying basic correlations between environment and marmot biology, the following review will be brief. Some of the clearest correlations are those between habitat and patterns of reproductive effort. Woodchucks experience the longest growing seasons of all marmot species, living in low-elevation fields and forest ecotones. They become sexually mature as yearlings, breed annually, and disperse as young of the year. Yellow-bellied marmots, inhabiting intermediate-elevation meadows in the Sierra Nevada, South Cascade, and Rocky Mountains, become sexually mature one year later than woodchucks, as two-year-olds, and they disperse one year later as well, as yearlings. They may breed annually, like woodchucks, but individuals often skip a reproductive year. Olympic and hoary marmots, inhabiting subalpine and alpine meadows at or above timberline in the Olympic, northern Rocky, and Cascade Mountains, are delayed yet another year: they do not become sexually mature until they are at least three-year-olds, they do not disperse until they are two-year-olds, and they never breed more frequently than in alternate years.

Although more studies are needed, it appears that some other species of marmots are consistent with this trend. Thus, the Vancouver marmot resembles the Olympic and hoary marmots in its basic sociobiology, as does the European Alpine marmot and the Brower's marmot, the latter limited to Alaska's Brooks Range, where high latitude compensates for relatively low altitude. The Eurasian bobak marmot, found in the steppes north of the Caucasus, appears to resemble *M. flaviventris*, whereas the other, high-elevation species show more similarities to *M. olympus* and *caligata*.

There are other correlations. Occupying one environmental extreme, woodchucks are the smallest marmot, while at the other extreme, Olympic and hoary marmots are the largest in North America. Yellow-bellied marmots are appropriately intermediate in body size and environment. Finally, woodchucks are the most solitary and socially intolerant of all marmots; Olympic and hoary marmots are the most interactively colonial and socially tolerant, and yellow-bellied marmots, once again, are intermediate. Something, it seems, is going on here. But what?

These fundamental trends were first recognized more than ten years ago (Barash, 1973c) and integrated into a general theory relating environmental regimes, age at sexual maturation and dispersal, and the patterning of social behavior (Barash, 1974a). Subsequent findings have led to modifications and adjustments, although the basic structure has been supported.

Marmots occupying low-elevation environments with a long growing season mature early and disperse one year before they mature. This is consistent with the general trend among vertebrates of dispersal preceding sexual maturation, and it suggests that the age at sexual maturation may be cued to the time required for young animals to achieve adult body

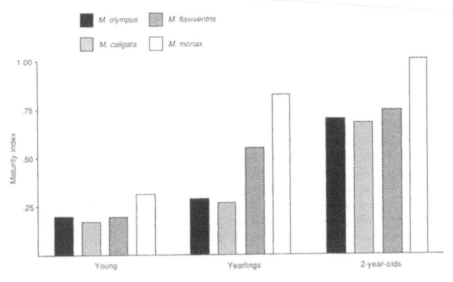

Fig. 14.1. Maturity indices for four species of North American marmots. See text for details.

size—which, in turn, seems related to the environments occupied. Thus, yellow-bellied marmots, inhabiting a more severe environment, delay sexual maturity one year later than do woodchucks, and their dispersal is correspondingly delayed as well. And Olympic and hoary marmots, in turn, delay their sexual maturation and their dispersal yet another year. A "maturity index" can be calculated for each species by dividing the mean weight of the age class under consideration by the mean weight of adults at the same time of year (Fig. 14.1).

Maturity indices reveal that as young of the year, *M. olympus, caligata,* and *flaviventris* average about 20% of adult weight; they typically do not disperse. By contrast, young of *M. monax* average greater than 30% of adult weight, and they disperse. One year later, yearling Olympic and hoary marmots have achieved about 30% of adult weight; they still do not disperse. By the time they are yearlings, however, yellow-bellied marmots have exceeded 50%, whereupon they disperse. Yearling woodchucks, already independent, and some of them reproducing as well, are about 80% of adult weight at this time. Another year later, two-year-old yellow-bellied marmots have reached nearly 75% of adult body weight, and some of them begin to breed; by contrast, two-year-old Olympic and hoary marmots are at less than 70% of adult weight. As two-year-olds, however, these high-elevation species finally disperse, delaying reproduction until at least the following year, when they are three-year-olds and typically at 90% of adult weight.

There is no across-the-genus threshold percentage of adult weight that must be achieved in order for reproduction to begin. Although woodchucks breed as yearlings when they are at about 75% of adult weight, hoary marmots do not breed when they have attained the same relative size: they reach 75% of adult weight as two-year-olds but do not breed until the following year and often another year after that. As environmental regimes are more severe for the latter than for the former, it is not surprising that for reproduction to occur, *M. caligata* must approach its adult weight more closely than must *M. monax*. Since relative annual growth rates are lower for *M. caligata*, as indicated by the differences in maturity indices, delayed sexual maturation is doubly indicated among the proportionately slower-growing species.

Dispersal is tightly coupled to age of sexual maturation, and it takes place at the appropriate and predicted ages in the different species of marmots. Because it influences dispersal and sexual maturation, the environment of each species also seems likely to influence the social system. Thus, selection should favor any social system that confers maximum reproductive success upon its members, including the dispersers. Aggressive behavior by the adults helps initiate dispersal among yearling yellow-bellied marmots, and as woodchucks are still more aggressive, a similar or even stronger correlation seems likely in that species. In fact, as we have seen, the greater aggressiveness of woodchucks may well be related to their dispersing at an earlier age. Dispersal in Olympic and hoary marmots is delayed even longer than in yellow-bellied marmots and, in fact, longer than in any other marmot species known. *M. olympus* and *caligata* are also the most colonial and socially tolerant species; indeed, nondispersal can be seen as both a proximate and ultimate cause of coloniality and social tolerance. (Alternatively, it could also be a *consequence* of coloniality and social tolerance.)

In any event, the delayed dispersal of the montane species, which is presumably made necessary by environmental constraints, large body size, and consequent low maturity indices, may well be made possible by a social system that is, by marmot standards, extraordinarily tolerant. Aggressiveness—as we have seen—is apparently related to the onset of dispersal, although in complex ways. Selection may therefore have favored different levels of aggressiveness in each environment, maintaining the competitive success of colony residents while also generating the optimum age at dispersal. The progressive increase in sociality among marmots experiencing progressively shorter growing seasons may therefore be due to the increasing necessity of inhibiting the dispersal of undersized animals.

Male/male competition, prominent among all marmots, is no less intense among the highly colonial forms than among the solitary. Variance in male reproductive success among the former species may be as high or even higher than among the latter, since Olympic and hoary marmots are more obviously polygynous. If, as seems likely, the montane forms prove to be more subject to polygyny, then delayed maturity may result in part from male/male competition as well. In other vertebrates, such a pattern leads to sexual bimaturism—i.e., delayed maturation of the more competitive sex (e.g., Wiley, 1974). Female/female competition, resulting from colonial living arrangements, should similarly retard sexual maturation among females. This scenario is problematic, however, as it assumes that coloniality generates intense intrasexual competition and then derives coloniality from the resulting delayed maturation and dispersal. It seems more likely that if intrasexual competition is involved in the evolution of coloniality via delayed maturation, it is contributory rather than a major cause in itself.

The intolerance among woodchucks, the corresponding tolerance among hoary and Olympic marmots, and the intermediate tolerance shown by yellow-bellied marmots may also be related to the differing frequency of transients encountered by these species. A population structure characterized by frequent interactions among unrelated individuals might be expected to be less altruistic, and vice versa. This effect might also be exaggerated if the population of nonrelatives also consists of many transients, thereby disfavoring the evolution of social cooperation via reciprocity (Trivers, 1971; Axelrod, 1984), and if the population of relatives is highly viscous, as in *M. caligata* and *olympus*. However, it is also possible—and in the case of marmots more likely—that causation works primarily the other way: a lower frequency of encounters with transients (among hoary and Olympic marmots) seems to be a partial *result* of social tolerance, rather than a cause, just as the relatively high frequency of transients among woodchucks seems to be a partial result of their social intolerance, which is correlated in turn with early dispersal in this species. Yellow-bellied marmots, once again, appear to be roughly intermediate.

In addition to age at maturation, age at dispersal, and social pattern, the structuring of reproductive effort among marmots appears to be influenced in a consistent way by natural selection, acting under consistently varying environmental constraints. There is a cline in reproductive effort, from the woodchuck (the highest among marmots) to the Olympic and hoary marmots (the lowest). There are several nonexclusive ways in which selection has probably produced this cline. For one, because of nutrient limitation and the constraints of long hibernation periods, metabolic demands are almost certainly most severe on the high elevation species, more relaxed on medium-elevation yellow-bellied marmots, and eventually "bottoming out" among the low-elevation woodchucks. This species, in addition to maturing and dispersing earliest, also reproduces the most often, expending the most reproductive effort.

Once coloniality has been established, kin selection could conceivably operate to retard reproductive rates, since colonies tend to be composed of close relatives. By reproducing less frequently, individuals minimize competition with replicates of their own genes. The apparent ability of marmots to regulate colony size within certain limits, most clearly developed among Olympic and hoary marmots, could also be an adaptive response to the constraints imposed by the occupation of fragile environments by long-lived, sedentary animals. However, although it seems likely that kin selection is involved in the patterning of social behavior among marmots, current evidence suggests that it has not been responsible for reduced reproductive efforts; if anything, the opposite seems to hold, since reproductive success is typically higher for individuals associated with relatives.

Additional Tests and Equivocations

These correlations and emerging theory have led to further research and numerous efforts at testing. For example, when I began my initial study of hoary marmots (in 1970) nothing was known of the sociobiology of this species except that it occupies subalpine and alpine environments similar to those of the Olympic marmot. Because of this environmental similarity, I predicted adaptations similar to those of *M. olympus*; this was confirmed (Barash, 1974c; Holmes, 1984b). Similar predictions were made and confirmed for the European Alpine marmot (*M. marmota*), as measured by physical proximity, and chasing and play-fighting frequencies, all of which revealed a high level of social tolerance and coloniality (Barash, 1976b). A laboratory study of the Brooks Range species, *M. broweri*, suggested that it too is socially tolerant and matures late (Loibl, 1983). On the other hand, frequencies of greeting behavior in *M. marmota*, employed previously as a convenient indicator of sociality among the North American marmots, were found to be consistently and significantly lower than in *M. olympus* or *caligata*. This finding suggests that many factors should be considered in comparing marmot "sociality," as the evolution of social behavior in this genus has apparently involved the independent elaboration of various distinct social characteristics.

Since yellow-bellied marmots occupy a range of environments, they were expected to exhibit a range of intraspecific adaptations analogous to those found interspecifically within the genus *Marmota*. Accordingly, I compared the sociobiology of two colonies of yellow-bellied marmots in Rocky Mountain National Park, Colorado: one colony at 2,650 meters and the other at 3,850 meters (Barash, 1973d). The social behavior of the medium-elevation colony resembled that previously described for the same species in comparable environments in Yellowstone National Park, Wyoming (Armitage, 1962), whereas the high-elevation colony demonstrated those characteristics described for the Olympic and hoary marmots.

Unknown to me at the time, a similar pattern of social tolerance among high-elevation yellow-bellied marmots had also been reported for a Montana population living above 3,000 meters, in which five to eight adults were often seen "feeding or basking in a single small area" (Pattie, 1967). This situation is found only rarely among medium-elevation colonies of this species. However, variability in social behavior is the rule rather than the exception for *M. flaviventris* (Armitage, 1977), and social behavior varies with many factors intrinsic to a particular colony's makeup at any given time: the number of reproductive females, their behavioral profiles, the numbers of transients and of returning individuals from the previous year, etc. Such factors introduce considerable variance into observed patterns; it remains to be seen whether this represents evolutionary "noise" superimposed on underlying regularity or whether it constitutes its own pattern independent of the genus-wide trends here suggested.

Comparisons of high- and medium-elevation yellow-bellied marmot colonies in Colorado have added to the cogency of the proposed intraspecies trends, although interpretations differ (Johns and Armitage, 1979). Whereas animals living at 2,900 meters occasionally breed annually, at 3,400 meters no adult female weaned young in consecutive years, and only one of four two-year-olds weaned a litter. In fact, the reproductive effort expended by high-elevation *M. flaviventris* (0.94 young per adult female per year) was comparable to that of *M. olympus* and *caligata*, and only half that for *flaviventris* at medium elevation (1.99: Armitage and Downhower, 1974).

Moreover, the alpine population of *M. flaviventris* revealed extensive overlap of home ranges, an average of 90% over three years. Amicable encounters were more frequent than agonistic encounters for all age and sex pairs, if avoidance of the adult male by the adult female is excluded. The high-elevation site also evidenced fewer agonistic interactions than did the medium-elevation site: in the former case, female yearlings in particular avoided encounters with individuals outside of their burrow group (Armitage and Johns, 1982). The high-elevation site also supported a higher density of animals—3.7 animals/hectare vs. 1.3 animals/ hectare at medium elevation. Since forage abundance *per se* did not appear to be limiting at either the high site (Andersen, Armitage, and Hoffmann, 1976) or the low one (Kilgore and Armitage, 1978), the higher density at higher elevations may well have been related to the diminution in agonistic behavior.

Alternatively, the observed differences in local population density may reflect the distribution of physical features, notably the presence of rocky outcrops suitable for burrow sites (K. B. Armitage, personal communication). Thus, the high-altitude colony had at least four such outcrops near good foraging meadows, whereas medium-elevation colonies usually had one or two. It is therefore conceivable—and a hypothesis worth exploring—that the observed patterns of social behavior are at least partly a result of "enforced togetherness," directly mediated by the natural environment. In any event, social adaptations are apparent, and they appear to have other correlates as well. For example, young of the year at the high-elevation site showed less variability in individual growth rates than did young at medium elevations (Andersen, Armitage, and Hoffmann, 1976), suggesting more equitable access to forage and, hence, less socially mediated exclusion.

Of particular interest is the observation that whereas 95% of yearling males and 54% of yearling females disperse from medium-elevation *M. flaviventris* colonies, at the high-elevation site only 76% of yearling males and 36% of yearling females dispersed. Thus, it appears that when yellow-bellied marmots occupy environments similar to those occupied by Olympic and hoary marmots, they display behavioral and biological traits that resemble those of the latter two species. It remains to be seen, of course, whether these adjustments represent adaptations in the evolutionary sense or direct phenotypic consequences of differing nutrient regimes, physiognomic factors, etc.

(It is also possible that this relationship—and others presented in this book—will be found to crumble when and if sample sizes are increased. Thus, the medium-elevation data discussed above are based on about 17 years of study, whereas the high-elevation data are based only on about three. As there is considerable year-to-year variation at the medium site, it is possible that the high-elevation data happened to be collected in atypical years. This problem, of course, arises with any field work—indeed, any biological research—but is particularly acute when one seeks to generalize across different species or across different populations of the same species.)

Regrettably, little information exists regarding the social behavior of the Palearctic marmots, largely because Soviet field biology tends to emphasize habitat description, anatomic and physiologic details, and such economically relevant factors as fur management, parasites, and marmots as disease reservoirs. Nonetheless, studies of the Palearctic marmot species consistently suggest correlations between climatic severity and nutrient regime; for example, Altai marmots (*M. baibacina*) feeding on lush, wet alpine meadows are reported to grow more quickly during the early summer than their steppe- or forest-dwelling counterparts. However, this changes dramatically as the season progresses, so that by emergence from hibernation, three populations of the same species evince the following percentages of body fat among adult males—alpine: 9.3%, steppe: 13.3%, and forest-meadow: 26.1% (Bibikov and Berendaev, 1978). Alpine populations of the long-tailed marmot are known to experience greater instability in early spring conditions, which in turn results in a higher proportion of embryonic resorption (Kizilov and Berendaev, 1978). It is not known whether these species also exemplify the behavioral adaptations that might be predicted, although it may be noteworthy that the "territories" of alpine long-tailed marmots are consistently larger than those of forest-dwelling populations of the same species (Davidov et al., 1978).

In proportion as differing reproductive and social patterns represent evolutionary adaptations, immigration may reduce the precision of any such adaptations, with the result that intraspecies correlations (and hence the adaptedness of individuals) would be less than interspecies correlations. Thus, of 62 medium-elevation yellow-bellied marmot litters whose emergence dates were recorded from 1962 to 1972, six emerged after July 21; of the 22 young in these litters, only two survived to be yearlings the following spring. It may be significant that not one of these apparently maladapted litters were produced by females that had been born in the study area: all were immigrants (Armitage, Downhower, and Svendsen, 1976). They may well have dispersed from higher-altitude populations, for which later emergence and later weaning are adaptive because the snowpack persists longer into the early summer.

It should be emphasized that we do not know the extent to which many life-history traits among marmots are evolved or simply phenotypic responses. Insofar as genetic adaptation is involved, however, it may be noteworthy that transients are more frequent among yellow-bellied than among Olympic or hoary marmots; accordingly, adaptive precision should be less in the former than in the latter. Transients are most frequent among woodchucks, but, on the other hand, the specificity of local adaptations is presumably also the least in this species, so there is less to be disrupted.

Marmots in Particular and Sciurids in General: Adaptive Patterns in Size, Severity, and Sociality

The proposed general theory of environment/behavior correlations emphasizes that varying nutrient regimes probably act via relative growth rates. Yet growth of young and yearling *M. flaviventris* at high elevation is no less than, and often exceeds, that of medium-elevation animals (Andersen, Armitage, and Hoffmann, 1976; Armitage, Downhower, and Svendsen, 1976), although the difference is not statistically significant. At the same time of the year, however, high-elevation individuals are consistently lighter; e.g., adult females in July averaged 2.32 kg at 3,400 meters vs. 2.58 kg at 2,900 meters; young of the year in August averaged 0.72 kg at 3,400 meters vs. 1.03 kg at 2,900 meters. Medium-elevation young weaned in early July weigh about 1.67 kg, whereas young at the high-elevation site would not reach the same weight until about September 15—if growth rates continue without diminution in August and September.

Even this assumption is unlikely to hold, given the conspicuous late-summer desiccation of most mountain meadows. Moreover, in order to achieve a body weight equal to that of 2,900-meter animals, 3,400-meter yellow-bellied marmots would have to maintain this growth rate well into October, by which time snowfalls are common.

Especially in a comparison of different populations of the same species, a higher *rate* of weight gain might in fact be expected in the population occupying a more severe environment, if a brief period of above-ground activity takes place in meadows that are lush although short-lived. If such a rapid rate continued for a period comparable to that enjoyed by the inhabitants of more moderate environments, then more weight would be gained by the former animals than by the latter. One can hardly conclude from *daily* growth rates, however, that elevation and length of growing season do not have a significant effect on growth processes, and a likely retarding one at that, unless yearly rates are considered. Moreover, larger overall size may well be required for success at higher elevations; therefore, yearly rate relative to adult body size becomes the most appropriate measure. This is strongly suggested by the progressively retarded dispersal ages in the sequence *monax, flaviventris, olympus/caligata*, as well as by the reduced reproductive rates also found in this same sequence. Montane yellow-bellied marmots from 2,900 meters in Colorado are also significantly heavier than low-elevation conspecifics from the Columbia River valley in eastern Washington State (Ward and Armitage, 1981b). This suggests that larger body size is adaptive at higher elevations, even though because of the shorter period of above-ground activity, it may be more difficult to attain. (Here again, a caveat is regrettably in order: since the age structure of the Washington population was unknown, it could have been biased toward younger individuals, which are also smaller.)

Even if nutrient abundance per day is higher at higher elevations— with this difference reflected in daily weight gains as well—the total number of available days becomes crucial, and it seems evident that higher-elevation populations have fewer foraging days available to them. On the other hand, since for marmots heat stress appears to be more severe than cold stress (at least during the summer), the number of hours available for foraging per day may be locally greater for higher-elevation populations. In any event, it is noteworthy that in one study, weight-gain data on a high-elevation *M. flaviventris* population were terminated on September 12 (Andersen, Armitage, and Hoffmann, 1976), following a snowstorm after which weight gain ceased. Given the likelihood of such autumnal snowstorms at high elevations, it seems unrealistic to postulate continued regular weight gain into mid- or late October.

In general, the young of hibernating sciurids grow more rapidly than do the young of nonhibernators (Maxwell and Morton, 1975), presumably because they have a shorter active period in which to do so. Moreover, inhabitants of severe climates grow faster than do inhabitants of more moderate climates (Kiell and Millar, 1978). It is not surprising, therefore, that high-altitude yellow-bellied marmots grow rapidly.

In response to the likely demands of a severe, high-elevation habitat, successful dispersal may well require growth that is faster than that at lower elevations. Dispersal is more likely to be delayed in the former case, and selection may even favor growth rates exceeding those of lower elevation, perhaps at the sacrifice of some other consideration, such as reduced predator avoidance because of greater foraging effort, etc. Such adaptations may nonetheless be insufficient to permit early dispersal; accordingly, they could select for a degree of social tolerance within the colony.

It seems clear that "growing season" is important to marmot sociobiology; it is also equally clear, unfortunately, that it is a difficult commodity to identify. Thus, ecologists typically use it to refer either to the frost-free period or the period the ground is free of snow. The latter appears to be especially important for marmots, and Andersen, Armitage, and Hoffmann (1976) proposed that the period free of major snowstorms corresponds best with the actual time of marmot weight gain. This also seems likely to be the critical measurement as far as the animals themselves are concerned, although the period of weight gain appears to be less important than the weight actually gained and, more to the point, the amount gained relative to the amount that must be gained. Thus, Olympic marmots gain weight for up to 120 days (Barash, 1973a), yellow-bellied marmots at high elevation, for only 96 days (Andersen, Armitage, and Hoffmann, 1976), although the latter nonetheless mature a year before the former.

The actual time available for foraging may in fact be somewhat less than the length of the snow-free period, since animals may be constrained from foraging by factors other than the simple availability of something to eat. For example, among yellow-bellied marmots at low elevations in the desert of eastern Oregon (Webb, 1981), foraging activity is independent of ambient temperature so long as it is below 25° C, a "critical temperature" at which marmots show a minimal metabolic rate (Kilgore, 1972). Above that temperature, marmots are heat-stressed; to avoid this stress, they remain in their burrows. Hence, elevation as such cannot be used as the *sine qua non* of marmot sociobiologic adaptations, since low-elevation populations may well experience substantial restrictions of their food supply.

For example, *M. bobak* lives at low elevations and consequently enjoys a long "growing season" on the Eurasian steppes. But the actual growing season as experienced by the animals is apparently quite short, since the young are born 1 to 2 weeks after the adults emerge in spring, which indicates that breeding occurs in the burrows before emergence, presumably because of a restricted above-ground activity period, even at their low elevation. Thus, in *M. bobak* habitat in Soviet Kazakhstan, the young emerge after the first week in May, and hibernation may begin as early as the end of July; this is a region regularly subject to severe summer droughts (Shubin, 1962, 1963). I have observed that a similar pattern of drought coupled with early hibernation occurs among low-elevation yellow-bellied marmots in the deserts of eastern Washington State.

Although such measures as the period of weight gain are preferable to "vegetative growing season," the former is nonetheless unsatisfactory, because it measures the animals' response to their environment rather than the environment itself. As marmots are presumably adapted to their environment, the former cannot be taken as an objective measure of the latter. Ideally, such an assessment should include "effective foraging time," corrected for day length as well as periods of heat stress (Webb, 1981), especially at lower elevations where growing season *per se* might be misleadingly high. But it should also include some indication of nutrient availability per foraging day, since lush subalpine forage—even if available for only a relatively short period—can provide substantial nourishment. In addition, even though lower elevations are characterized by longer growing seasons and milder temperatures, they may offer less food value simply because xeric habitats typically have sparse forage low in calories.

If food plants are widely separated, then search time could also become an issue. Otherwise, with sufficient forage, gut clearance rate is likely to be a limiting factor in marmot weight gain, ultimately influencing maturation rate and hence age at dispersal, reproductive rate, and social tolerance. A more inclusive measure should also include avoidance of predators, since predator density and the availability of safe refuges and surveillance sites presumably influence available above-ground foraging time.

Following the appearance of a predator at a colony of *M. olympus* or *caligata*, the animals typically remain within their burrows, sometimes for many hours. Thus, the appearance of a cougar (*Felis concolor*) at a hoary marmot colony at 1800 hrs in early July terminated all above-ground activity for the day, and no animals emerged until after 0800 hrs the following morning as well, more than 2 hours later than is typical. Although there is no evidence that successful predation is more likely to inhibit above-ground activity than is the mere appearance of a predator, it seems clear that in addition to killing directly, predators can greatly restrict foraging time as well. (It remains to be seen, however, whether chronic encounters with predators eventually induce marmots to resume their typical foraging patterns.) In any event, just as we might ultimately hope for a multidimensional scale for the evaluation of social behavior, we might similarly recognize the desirability of a multidimensional scale for the assessment of "environmental severity."

Even though we currently lack a single, bottom-line inclusive measure of this sort, analogous to our current understanding of inclusive fitness, it seems nonetheless clear that some significant correlations between marmot biology and marmot environments exist. We are also left, however, with a widely encountered paradox: in a congenial environment, a population of marmots should increase, as a result of which socially mediated competition should also increase. And this, in itself, would generate an increase in "environmental harshness."

If so, then the various extrinsically imposed measures of environmental harshness experienced by different marmot species are themselves not meaningful measures of anything, since individuals of each species experience their own regime, perhaps balancing density-independent and density-dependent factors in a unique way, resulting in an overall level of environmental severity which is roughly similar for all species within the genus. The outcome of such a balancing act among factors of environmental severity may well resemble an "ideal free" distribution (Fretwell and Lucas, 1969), in which individuals of each species experience environments that are measurably different in a variety of ways but each of which may be equally "severe" and within which each species can be equally "successful."

But even if this were so, the patterns of biological, reproductive, and behavioral adaptations by which this "equality" is achieved would nonetheless differ among species; once again, we are justified in searching for consistent and potentially revealing patterns.

Dispersal seems particularly relevant in this respect, especially with regard to its apparent relationship to reproductive rates, age, size at maturity, and patterns of social tolerance. Late dispersal is at least partly a response to reduced growth rates relative to adult size and delayed sexual maturity. This in turn is characteristic of marmots occupying environments such as high-elevation mountain meadows, with likely consequences for coloniality and social tolerance. It has also been suggested (Johns and Armitage, 1979) that the observed biennial breeding of high-elevation yellow-bellied marmots itself *permits* delayed dispersal, since when mothers are not pregnant, they are less aggressive and therefore less likely to drive away their young of the previous year. By remaining as nondispersers—or, at least, delaying their dispersal for an additional year as in *M. olympus* or *caligata*—these offspring are less likely to interfere with their mother's next reproduction than if the mother were reproducing annually. For her part, the mother is also able to invest in her offspring for at least an additional year. And by garnering one more year of growth before venturing away from their natal colony, offspring increase the chances of their ultimate reproductive success. Assume that delayed sexual maturation is imposed by the extrinsic environmental regime and the need to attain a minimal body size, which in montane species is quite large. Then by delaying their dispersal, such animals are not so much delaying their reproduction as increasing the chances that they will survive until they are large enough to be able to reproduce.

Among the ground squirrels generally, proximate causes of dispersal are difficult to identify. Direct aggression, notably from the mother, has not been implicated in the dispersal of males (*Spermophilus beldingi*: Holecamp, 1983, 1986; *richardsonii*: Michener, 1973; *tereticaudus*: Dunford, 1977b; and *beecheyi*: Dobson, 1979). On the other hand, aggression may be influential in the dispersal of females (*S. elegans*: Pfeifer, 1982), in avoidance (Yeaton, 1972), and in adults' failure to respond to the overtures of juveniles (Michener and Sheppard, 1972). Juvenile female Columbian ground squirrels greeted their mothers three times more frequently than did juvenile males (Waterman, 1985), a difference that may be associated with male-biased dispersal. In at least one species, dispersal seems more frequent among juveniles from large than from small litters (*S. tereticaudus*: Dunford, 1977b).

Whereas the mother's disappearance did not influence dispersal of *S. beldingi* (Sherman, 1976), yearling males of *S. richardsonii* are likely to disperse farther when their mother survives (Michener and Michener, 1973). Dispersal is not distinctly density-dependent in *S. elegans* (Pfeifer, 1980), *beecheyi* (Dobson, 1979), *armatus* (Slade and Balph, 1974), or *beldingi* (Holecamp, 1983); however, dispersal of *S. richardsonii* tends to be low when population density is low (Michener, 1979). The termination of dispersal, at least for females, shows a distinct tendency to be density-dependent, in that a lower population of conspecific females tends to correlate with settling (*S. parryi*: Green, 1977; *S. armatus*: Slade and Balph, 1974; and *S. beecheyi*: Dobson, 1983). These observations are basically consistent with the marmot pattern, notably the roles of resource limitations and female/female competition.

As to the adaptive significance of sciurid dispersal, three major possibilities may be mentioned: resource competition, inbreeding avoidance, and enhanced access to mates. As among marmots, resource competition among the smaller sciurids seems to be more influential on females than on males (see review in Holecamp, 1984). Inbreeding avoidance appears to correlate with male-biased dispersal, as observed in *Cynomys ludovicianus* (Hoogland, 1982), *S. beecheyi* (Dobson, 1979), *S. elegans* (Pfeifer, 1982), *S. richardsonii* (Michener and Michener, 1977), and *S. beldingi* (Holecamp, 1983, 1986). Enhanced access to mates appears to influence males' dispersal in particular, and removal of adult males tends to inhibit juveniles' dispersal in both *M. flaviventris* (Brody and Armitage, 1985) and *caligata*. In most cases, resource competition, inbreeding avoidance, and access to mates tend to covary, making it difficult to attach particular importance to any one.

Whatever the proximate and ultimate factors related to dispersal in general, the exact role of *delayed* dispersal among marmots occupying higher elevations also remains uncertain, although the correlation is now well established. Whether dispersal occurs early or late, its significance with regard to resource competition, inbreeding avoidance, and mate acquisition remains roughly unchanged, so long as it occurs prior to sexual maturation, which it does. Perhaps delayed maturation and dispersal drive the high-elevation system of reduced reproductive rates and greater social tolerance. Or, alternatively, perhaps it is an independent adaptation to the same environmental factors as the latter. Regardless of its precise causative sequence, the system itself is clearly robust, both interspecifically and to a lesser extent intraspecifically.

As we have seen, adult body size also varies among marmots, both interspecifically and to a lesser extent intraspecifically. Body size seems to be significant both as a dependent variable—an adaptation to environmental regimes—and as an independent variable, a "goal" to be attained by young animals, and one that requires various other adjustments. The pattern seems to go beyond the genus *Marmota*. Variations in sociality among the ground squirrels in general correlate closely with age at first reproduction as well as the age at which adult weight is attained, which in turn correlates with growing season (Armitage, 1981b) as suggested for the marmots (Barash, 1974a).

Among the ground squirrels, sociality may have evolved at least partly because of the benefits of retaining daughters beyond weaning, combined with the development of affiliative bonds among adult females and the association of adult males with the female group (Michener, 1983). As a general rule for all animals, the overlapping of generations tends to correlate with the evolution of more elaborate social systems, and the greater the degree of iteroparity, the more likely are overlapping generations. Larger, slower-growing, and slower-maturing species are more likely, in turn, to be iteroparous and to experience overlapping generations and well developed sociality. Michener (1983, 1984) also emphasized a correspondence between sociality and the overlap of adults' and juveniles' above-ground activity: among the ground squirrels generally, species in which the above-ground seasons of adults and juveniles overlap less tend to be relatively asocial, and vice versa. Yet even the more solitary species form kin clusters in which sisters and/or mothers and daughters live close together. This seems to provide opportunity for sociality to develop, if other selective pressures are conducive. Among the marmots, moreover, adults and juveniles of *M. monax* overlap seasonally as much as those of *M. flaviventris* and *caligata*, yet woodchucks are substantially less social.

The correspondence between sociality and adult/juvenile above-ground activity overlap may occur because the former causes the latter rather than the other way around. One aspect of sociality, in fact, is temporal and spatial overlap in the activity patterns of adults and juveniles. I suggest, moreover, that sociality and adult/juvenile activity tend to coincide largely because each is an independent response to the same environmental selective factor(s), related primarily to nutrient availability, environmental harshness, maturation factors, and size energetics, with an additional assist from kin selection and sexual selection. If animals mature and disperse late, they remain in their natal colonies longer and get to overlap temporally with adults, especially if this late development is a response to nutrient regimes (and size energetics) that keep adults active above ground for as long as possible. Activity coincidence may serve as a proximal mechanism facilitating adult/subadult bonding, occurring as a result of the ultimate factors causing sociality, but almost certainly not as an ultimate cause in itself.

Iteroparity is most pronounced among large-bodied, slow-growing, and late-maturing species, notably the marmots among the sciurids, and the montane species among the marmots. Delayed maturity, in turn, results from restricted food availability (including a shortened growing season at higher elevations) and larger adult body size, a goal that obviously takes longer to reach in larger species. Among species with large-bodied adults, maturity index tends to be low, and as we have seen, this correlates with delayed maturation, delayed dispersal, and appropriately varying environmental regimes and social systems.

In the sociobiology of the other ground squirrels, numerous correlations are apparent, although none are as clearcut as those in the genus *Marmota*. Both Armitage (1981b) and Michener (1983) identified five levels of sciurid sociality, a characterization that helps order an otherwise confusing group of animals. The scheme presented here, despite some minor modifications, is basically consistent with theirs. In some cases, species are located arbitrarily (e.g., *S. tridecemlineatus* could be placed in either the "asocial" or in the "mother/offspring cluster" level). Intraspecific variation may also result in a species being incorrectly categorized simply because it is known from studies of only a limited number of populations.

The different levels of sciurid sociality can be identified as follows: (1) "asocial," in which the animals are essentially solitary, adults not associating together except for mating; (2) "mother/offspring clusters," within which mothers and offspring (notably daughters) tend to associate beyond weaning but between which interactions are rare and aggressive, with adult males not involved beyond breeding; (3) "male defense of female clusters," in which the adult females continue to live separately but an adult male defends a subset of females within his territory and juveniles from different litters may associate together; (4) "harems," in which one or more adult females and their offspring live together, intermingling with varying frequency, within the territory of an adult male; and (5) "integrated harems," in which cohesion within the harem is extreme and spatial and social restrictiveness is not clearly expressed.

The larger species—the marmots and prairie dogs—mature later and are more social than the smaller ones, the *Spermophilus* ground squirrels. This trend even holds to some degree within the genus *Spermophilus* itself: larger-bodied species such as *S. parryi*, *beecheyi*, and *columbianus* are the most social, whereas smaller-bodied species such as *tridecemlineatus*, *townsendii*, and *lateralis* are the most asocial, with the intermediate-sized *tereticaudus*, *beldingi*, *elegens*, *richardsonii*, and *armatus* intermediate in social system as well (Fig. 14.2). However, even this correlation is not perfect, since the most asocial ground squirrel, *S. franklinii*, is nearly twice the size of some of its more social congeners (e.g., *S. tereticaudus*) and the most asocial marmot, *M. monax*, is more than ten times the size of the moderately social spermophilines. The prairie dogs *Cynomys leucurus*, *gunnisoni*, and *ludovicianus* are all about the same size, yet they differ in level of sociality; in general, however, *Cynomys* is both larger than *Spermophilus* and more social. Size/sociality correlations are somewhat more reliable among the marmots, as are the correlations between delayed dispersal, delayed maturation, and reduced annual reproductive effort and sociality.

Given its apparent significance to ground squirrel and notably marmot sociobiology, the phenomenon of body size merits further attention. There seem to be many advantages of large size, including greater ability to survive cold weather, because of more favorable mass:surface area ratio. Bergmann's well-known rule states that "races from cooler climates tend to be larger in species of warm-blooded vertebrates than races living in warmer climates" (Mayr, 1963). On the other hand, absolute surface area increases with increased body size, even though relative surface area decreases, and absolute heat loss among mammals tends to increase with body weight to the 1/2 power (McNab, 1971). The insulating qualities of mammalian fur and skin may also be more significant than surface area/volume considerations (Scholander, 1955).

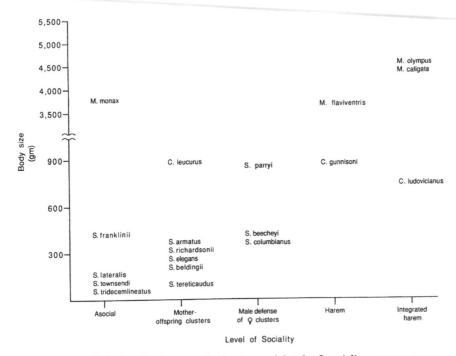

Fig. 14.2. Relationship between body size and level of sociality among some well-studied North American ground squirrels. Body size is presented as the mean weight (between emergence and immergence) of adult females.

Bergmann's Rule may have other adaptive bases. For example, seasonality is correlated with higher latitudes, and large animals, by virtue of their greater metabolic reserves and lower basal metabolic rate, may be more able to survive seasonal stress (Boyce, 1979), especially periods of acute food shortage (Searcy, 1980). For marmots, this may confer a particular advantage in successful hibernation, since larger animals can store more fat (Morrison, 1960) and dissipate it more slowly, thereby enjoying additional benefits in successful hibernation. In addition, size may contribute to a marmot's ability to survive inhospitable situations while seeking to establish a new colony or to gain admittance to an existing one.

Larger bodies also confer on a marmot greater ability to defend itself against moderate-sized predators such as fishers, martens, or badgers, and make it more likely to exceed the size avian predators such as red-tailed hawks can handle. Running speed tends to be greater among larger animals, and the energetic cost of running, per kilogram, tends to be lower; as a result, larger animals can maintain higher speeds for longer times. In *S. beldingi*, larger males also emerge earlier (French, 1982) and may well be reproductively more successful as a result. Females of *S. armatus* that emerge early tend to breed early and are more aggressive and successful (Balph, 1984); it seems likely that adequate size is a prerequisite.

Larger animals can exploit a wider range of forage, since with increased body size, gut capacity increases and relative protein needs decline (Janis, 1976). Thus, unlike the smaller sciurids, marmots can eat a diet composed almost entirely of stems, shoots, and leaves. Since coarse foods such as these are more evenly distributed as well as more abundant than the more calorically dense foods consumed by spermophilines, the larger body size of marmots may render them less subject to forage competition and thus more able to aggregate than the smaller ground squirrels. And once aggregation is possible, additional benefits can be accrued, via predator detection, kin selection, etc.

On the debit side, however, larger animals are easier to detect and less maneuverable, and they may be less able to exploit a range of possible burrows and crevices. Adult males of *S. richardsonii*, for example, become obese by late June, whereupon they immerge; at this time, they are less able to fit down refuge burrows and may be more susceptible to predation (Michener, 1983). It seems likely that all ground squirrels—including, presumably, marmots—immerge as early as possible, as soon as they attain adequate size and fat. Larger animals may also require more absolute quantities of food, and this in conjunction with the lower nutrient density of their forage typically requires that they spend more time foraging, during which they are more susceptible to predation and thermal stress. Most important, larger animals typically show a syndrome of lesser reproductive effort and, accordingly, lower reproductive rates. Each of the following increases with body size: gestation length, weaning age, age at sexual maturity, and interbirth interval. Litter size also decreases (Fig. 14.3), along with birth rate and the intrinsic rate of natural increase (see review in Clutton-Brock and Harvey, 1983). Larger mothers produce relatively smaller offspring, a trend throughout the Sciuridae as well (Armitage, 1981b); but these offspring, although relatively smaller, are absolutely larger, as in many other species (Ralls, 1976).

Sexual dimorphism among many vertebrates correlates with sexual selection, and male sciurids tend to be about 1.25 times larger than females (Armitage, 1981b). All male ground squirrels are aggressive during the breeding season, especially toward other males; among the asocial species, however, males may become subordinate to females following breeding, and there is no correlation between increased male role in the

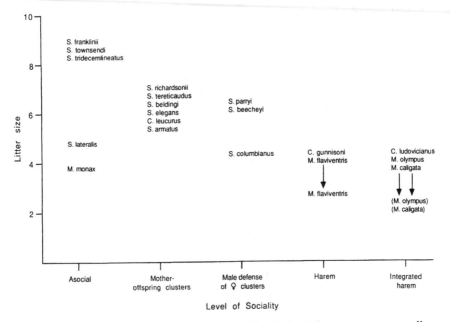

Fig. 14.3. Relationship between litter size and level of sociality among some well-studied North American ground squirrels.

social system and increased sexual dimorphism (Michener, 1983). It seems likely that within the size range of each species, large body size is selectively advantageous for both sexes, not only for overwinter survival but also because both males and females rely on accumulated fat during the critical weeks after spring emergence: males for the rigors of male/male competition, and females for breeding (e.g., Andersen, Armitage, and Hoffmann, 1976).

Nourishment is also critical to reproductive success among the smaller ground squirrels: *S. richardsonii* (Sheppard, 1972) and *tereticaudus* (Reynolds and Turkowski, 1972) produce more offspring in years of good forage, and when food is abundant, *Cynomys ludovicianus* breeds as a yearling (Koford, 1958). Late snowmelt has been found to correlate with reduced or failed breeding among several species of ground squirrels, including *S. columbianus* (Murie and Harris, 1982) and *lateralis* (M. T. Bronson, 1979). Following a severe winter, only 19% of female *S. lateralis* were pregnant (Phillips, 1984). And when their food was supplemented, a population of *S. columbianus* increased greatly, because of a combination of immigration, higher survival of resident young, and reduced age of sexual maturation among breeding females (Dobson and Kjelgaard, 1985a, 1985b). These results also suggest a fair degree of phenotypic plasticity in basic life-history characteristics.

Sociality among the ground squirrels correlates not only with body size but also with age of dispersal and age of reproduction; dispersal consistently occurs one year before sexual maturation. It is noteworthy that the most social spermophiline, *S. columbianus*, is the only one that delays dispersal until a yearling, and does not become sexually mature until two years old. All the other members of the genus *Spermophilus* mature as yearlings and disperse as young of the year (like woodchucks). The other two spermophilines classified at sociality level 3 (male defense of female clusters) are *S. parryi* and *beecheyi*. Significantly, the former takes advantage of the long arctic summer days and is therefore able to grow more rapidly during its juvenile year than any of its congeners (Mayer and Roche, 1954); otherwise, it would likely require an additional year to mature and disperse, like *S. columbianus*. *S. beecheyi* is on the verge of being assigned to sociality level 2 (mother/offspring clusters).

Among the prairie dogs, *C. ludovicianus* is the most highly social, and it disperses as a yearling and matures as a two-year-old. *C. leucurus*, by contrast, disperses as young of the year and matures as a yearling, like most of its spermophiline relatives. They may have long been associated with an environment that was overgrazed by bison and was therefore nutritionally poor (Koford, 1958), leading in turn to delayed dispersal and sociality. The situation among *C. gunnisoni* is unclear but suggestive: Raynor (1985) compared two sites supporting populations of *C. gunnisoni*. The two sites differed in availability of water, length of growing season, and abundance of vegetation. At the poorer site, the prairie dogs weighed less, reached sexual maturity later, and dispersed later. These correlations among dispersal, sexual maturation, and the age at which immatures reach adult weight are reminiscent of the situation among marmots.

The evolution of sociality among marmots seems, on balance, to result from numerous factors, of which body size is particularly important; body size itself seems to derive in a complex, nonlinear way from environmental regimes. Delayed dispersal is a necessary consequence of both large body size and delayed sexual maturation, but it is also a means whereby females are able to prolong their parental investment. Among most marmots, this maternal investment is female-biased, resulting in preferential recruitment of female offspring and dispersal of males. Females appear to clump in response to the concentration of environmental resources such as food and hibernacula. There is no evidence that females choose their mates; hence the "polygyny threshold" model of sociality (Orians, 1969) does not seem to apply to marmots, or, indeed, to any ground squirrels.

Among the more social spermophilines—e.g., *S. beecheyi* (Dobson, 1983)—males may guard females briefly following copulations, whereas among the more solitary species—e.g., *S. tridecemlineatus* (Schwagmeyer and Brown, 1983)—males simply wander and engage in scramble competition for estrous females. Multiple matings have been reported for several species, including *S. tridecemlineatus* (Schwagmeyer, 1984), *S. beldingi* (Hanken and Sherman, 1981), and *C. ludovicianus* (Foltz and Hoogland, 1981). The advantage of multiple matings to females, if any, remains to be determined. Multiple matings seem necessarily to be a disadvantage for otherwise dominant and highly fit males but an advantage for otherwise less successful males, since whatever its effect on females' fitness, multiple mating by females should reduce the variance in males' reproductive success.

When females are clumped and males' territories overlap such female clusters, males may be more likely to mate with females within their territories (e.g., as in *S. columbianus*, Murie and Harris, 1978), although multiple mating is a complication. To some extent, however, the male tendency to defend females correlates inversely with local population density: nondefense is more likely to be found at higher densities, defense at lower densities (Dobson, 1984). This may occur because the costs of herding, sequestering, and/or defending a harem become excessive when densities of both males and females are too high. Such a pattern is consistent with the traditional ethological wisdom that territoriality at low densities gives way to dominance hierarchies as density increases (D. E. Davis, 1958). A dispersed distribution of females of the relatively asocial *S. tridecemlineatus* appears to correlate with scramble competition among males, with male fitness being a function of their competence at locating mates (Schwagmeyer and Woontner, 1986). In this case, fluctuations in the density of males have little impact on levels of overt male/male conflict and competition, whereas changes in the density and breeding synchrony of females are more influential (Schwagmeyer and Woontner, 1985).

On the other hand, a clumped distribution of females could lead to highly efficient defense of females by males capable of doing so successfully, if the clumps of females are sufficiently separated from each other and the population density is not too high. This is especially true for the montane marmots, which although locally abundant are sparsely distributed over wider geographic areas and typically occupy discontinuous habitats. It seems likely that among the ground squirrels in general and the marmots in particular, clumping of females has in such cases led to prolonged associations of adult males with females, i.e., the maintenance of harems. Mountain meadows offer habitats that lend themselves to efficient patrolling and exclusion of unsuccessful competitors. Combined with relatively intense male/male competition and clumping of females, such systems should make the defense of harems especially adaptive for the males, facilitating sexual access and hence increasing reproductive success. For female prairie dogs, social grouping of this sort provides some protection from predators (Hoogland, 1981a, 1981b), and also facilitates the disbursement and receipt of kin-selected benefits (Hoogland, 1983).

Alarm calling tends to be kin-directed among the social species (Sherman, 1977; Dunford, 1977a), although self-preservation is also important for certain kinds of calls (Sherman, 1985). In *S. parryi*, close kin such as littermate sisters, or mothers and daughters, occasionally rear offspring in a common burrow (McLean, 1982), at least during the last few days before emergence. And related adult females of *S. columbianus* are more amicable than are distantly related or unrelated females, and they are also more likely to share resources such as food and burrows (Michener, 1983). In *S. beldingi*, close relatives are likely to support each other in agonistic encounters and to tolerate spatial encroachment (Sherman, 1980). Even in such asocial species as *S. tridecemlineatus*, behavior, notably in patterns of alarm calling (Schwagmeyer, 1980) and spatial overlap (Vestal and McCarley, 1984), tends to be kin-biased. In *S. richardsonii*, kin tend to displace each other less often than do nonkin (L. S. Davis, 1984), yet the species is primarily asocial. And for nearly all ground squirrels, regardless of level of sociality, spatial patterning is such that female kin in particular tend either to overlap or to live adjacently: *S. parryi* (McLean, 1982), *tereticaudus* (Dunford, 1977b), *richardsonii* (Michener, 1979), *beldingi* (Sherman, 1981), *columbianus* (Festa-Bianchet, 1981; King and Murie, 1985), *franklinii* (Haggerty, 1968), *elegans* (Pfeifer, 1982), *armatus* (Slade and Balph, 1974), and *beecheyi* (Dobson, 1979). At least some sciurids possess a finely tuned ability to discriminate relatives from nonrelatives, even at the level of paternal half-siblings, with whom they have never previously interacted (Holmes, 1986).

Such capacities and living arrangements facilitate kin-biased behavior (i.e., nepotism), but they do not require it. For example, it appears that among the sciurids, at least, when personal, selfish benefit—and its direct Darwinian fitness component—conflicts strongly with an indirect fitness benefit potentially available via kin selection, the former wins.

Thus, infanticide is a major source of mortality among black-tailed prairie dogs (*C. ludovicianus*), with the most common killers being resident lactating females, the most common victims the offspring of close kin (Hoogland, 1985). Neither males nor females of this species are more amicable toward close kin than toward more distant relatives (recalling my results with hoary marmots; see Chapter 10); moreover, nepotism on the part of males is generally reduced by competition among relatives for females. Similarly, nepotism among females is reduced by direct female/female competition for nesting burrows or breeding rights (Hoogland, 1986). It should also be noted that nepotism itself suggests but does not prove that kin selection is the prime selecting agent. And once again, a distinction should be made between kin selection and altruism.

The evolution of social behavior via female/female affiliations and kin selection requires that females be sufficiently sedentary to provide for interactions among relatives, and that they survive long enough for such interactions to occur and to be reproductively meaningful. Sherman (1980) has emphasized the role of demography in establishing possible "limits to nepotism" in *S. beldingi*. Two fundamental patterns of dispersal and mortality, both of which appear to be widespread among the ground squirrels, contribute to this effect. First, dispersal is generally male-biased, and second, because of intrasexual competition, mortality is also male-biased, even beyond the asymmetry introduced by dispersal itself. Both male-biased dispersal and male-biased mortality have been demonstrated for *S. beldingi* (Sherman and Morton, 1984). The sex ratio of adult *S. columbianus* is similarly skewed (Boag and Murie, 1981), and the risk-prone, wide-ranging behavior of male *S. richardsonii*—due presumably to mating efforts—has been directly implicated in higher male mortality in that species (Schmutz, Boag, and Schmutz, 1979).

On the other hand, sexual asymmetries in mortality are not universal among the sciurids. In both *S. franklinii* (Murie, 1973) and *M. monax*—both of them the least social of their genera—females and males disappear at approximately equal rates. It is tempting to suggest that male-biased dispersal and mortality among the ground squirrels correlate with sociality, in two likely ways. First, polygyny selects for more intense male/male competition, which leads to higher male mortality both because of fighting and high-risk, male-biased dispersal. And second, disappearance of males leaves the relatively long-lived females physically associated together, with greater opportunity to cohere socially via kin selection. (It should also be pointed out, however, that future research may yet reveal a high level of intrasexual competition and high male mortality even among the relatively asocial sciurids such as *S. tridecemlineatus*, *townsendii*, and *lateralis*, in which the breeding season appears to be characterized by intense scramble competition.)

Whatever the role of demography *per se*, nepotism and kin selection seem insufficient by themselves to select for sociality, since all of the sciurids have at least the potential for kin selection, yet many are not especially social. In proportion as a species is social, however, opportunities exist for individuals to enhance both their direct fitness (by entering into mutualistic, beneficial associations) and their inclusive fitness, promoting sociality still further. Moreover, body size seems likely to contribute further to this feedback phenomenon: in *S. beldingi*, for example, the low probability that distant relatives will interact limits the capability of nepotism, kin selection, and, presumably, sociality (Sherman, 1980). *S. beldingi* is a relatively small-bodied ground squirrel. Since body size and longevity are positively correlated, large-bodied species live longer and should therefore be more likely to interact and to have the opportunity of giving and receiving kin-selected benefits. Large body size accordingly seems likely to lead to sociality, not only through delayed dispersal and maturation, but also via kin selection.

But just as there are costs to large body size, there are also costs to sociality. Insofar as social grouping and especially harem organization tend to increase the variance in males' reproductive success, it imposes a cost on the unsuccessful peripheral or satellite male, not to mention the failed disperser. Whereas males most commonly bear the cost of being *excluded* from a social group, and of competition to enter such groups, females particularly bear whatever costs are associated with being *included* in such a group, since females constitute harems. Disease and parasite transmission, as well as increased aggression, have been suggested as the most important of these costs (Hoogland, 1979), along with inbreeding and the need to avoid it (Hoogland, 1982). It remains to be demonstrated, however, whether these costs actually are reflected in diminished reproductive success and/or inclusive fitness—and if they are not, in what sense such factors are "costly."

I suggest that the most important cost, one that until recently has not been adequately recognized by sociobiologists, is social suppression of reproduction. Among the spermophilines, for example, nearly 100% of all adult females reproduce every year; the notable exception, significantly, is *S. columbianus*, the most social species of this genus, in which the reproductive rate may be as low as 65% (Festa-Bianchet, 1981; Murie and Harris, 1982). Among the prairie dogs, *C. ludovicianus*—the most social—also appears to have the lowest percentage of breeding adult females. As we have already seen, reproduction is essentially 100% among adult female woodchucks, whereas social suppression of reproduction appears to be increasingly significant in the sequence from that species through *M. flaviventris* to *M. caligata*. It is always possible, of course, that social suppression of reproduction is mimicked in part by the syndrome of delayed maturation and relaxation of annually patterned reproductive effort that is characteristic of large-bodied, social ground squirrels generally. If so, then suppression of reproduction would be less definitely a cost of sociality than a component of the many factors responsible for sociality as an adaptive evolutionary strategy.

As already noted, even the normally tolerant hoary and Olympic marmots become more aggressive during pregnancy and lactation, suggesting that female/female competition is one of the costs of sociality. Many of the smaller sciurids reveal a pattern that can be interpreted as similar to the marmots', with overlap in home ranges being minimal during pregnancy and lactation: *S. parryi* (McLean, 1982), *columbianus* (Festa-Bianchet and Boag, 1982), *beldingi* (Sherman, 1981), and *S. tereticaudus* (Dunford, 1977b). Significantly, such trends are not discernible among the asocial sciurids, including *S. tridecemlineatus* (Wistrand, 1974) and *M. monax* (personal observations). Sociality results in female/female associations, which in turn seem to require a degree of aggressiveness when these females become reproductive. It may well be that aggressiveness is positively associated with parturition among the more asocial species also, but is simply less apparent because of their scattered distribution.

It should also be noted that social suppression of reproduction is not limited to females: following extensive removal of the resident adults, yearling males of *S. armatus*, which typically do not reproduce until they are two-year-olds, developed enlarged and descended testes, and some of them bred. At the same time, the percentage of reproducing yearling females increased (Slade and Balph, 1974). I do not know any circumstances in which male marmots have been observed to become sexually mature before their usual age of maturation.

Among some ground squirrels, notably *S. parryi* (McLean, 1983), males are frequently infanticidal, and male territoriality may well have evolved as a means whereby fathers protect their offspring, no less than a means of insuring continued access to estrous females. Among other species, notably *S. beldingi*, females are likely to be infanticidal (Sherman, 1981). In the former case, sociality—the progression from mother-offspring clusters to male defense of these clusters—may be an adaptive response to infanticide, whereas in the latter, sociality may impose a cost on the fitness of participating individuals if by virtue of their aggregation they are more susceptible to infanticidal neighbors. The role of infan-

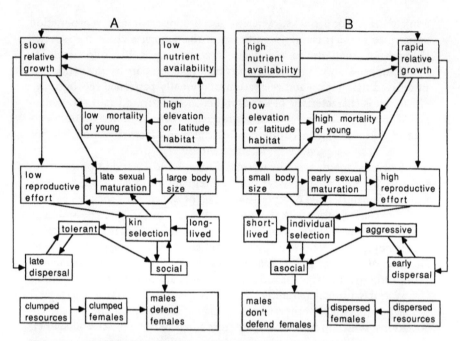

Fig. 14.4. Hypothetical flowchart of the interrelated factors selecting for a high level of sociality (system A) and a low level of sociality (system B) within the genus *Marmota*. The two systems are essentially mirror images of causality.

ticide in marmot sociobiology remains to be clarified, although it does not appear to be prominent in the genus.

Informed speculations about the evolution of sociality among various taxonomic groups often focus on a single, social modification during phylogeny. For example, it has been suggested that canid social evolution involved modification and elaboration of the male/female bond, whereas felid evolution revolved primarily around cooperating females (Kleiman and Eisenberg, 1973). The evolution of sociality among marmots (Barash, 1974a)—and perhaps among the sciurids in general (Armitage, 1981b)—apparently revolves around a different major axis: extended maturation and delayed dispersal of offspring. In addition, two other factors appear to be especially significant: affiliative behavior among adult females and prolonged temporal and spatial association of adult males with a harem. Other important factors are body size and nutrient regime in extended maturation and delayed dispersal, kin selection in female/female affiliations, and sexual selection in harem defense and maintenance.

Marmots offer a challenging and occasionally gratifying system that relates organism, environment, and social behavior via natural selection. The connections, however, constitute not so much a linear sequence of causes and effects as a feltwork of mutually supporting, sometimes confounding, and occasionally even contradictory adaptations and evolutionary sequences, which nonetheless have sufficient coherence to suggest patterns in the sociobiology of one genus (Fig. 14.4).

Other Systems and Some Predictions

The above considerations of ground squirrel sociobiology reveal that generalizations become increasingly unreliable for larger taxonomic units: patterns visible within the genera *Marmota*, *Cynomys*, or *Spermophilus* become increasingly dim when all three genera are combined in the search for patterns among the larger group. However, our confidence in generalizations that hold in one genus can be enhanced when the same generalization appears to hold in another (e.g., correlations between body size and sociality and between age of dispersal and sociality). Some intraspecies comparisons also offer suggestive parallels to the marmot case. For example, among golden-mantled ground squirrels (*Spermophilus lateralis*), populations found at higher elevations are characterized by shorter above-ground periods of summer activity, later age at first reproduction, higher survivorship of adult females, lower ovulation rates, smaller litter sizes, and heavier adult body weights (M. T. Bronson, 1979). And a population of *S. beldingi* at 2,246 meters elevation dispersed somewhat earlier than did individuals of the same species living at 3,050 meters (Holecamp, 1984). In both cases, unfortunately, possible effects on sociality remain to be determined.

In low-density populations, montane voles (*Microtus montanus montanus*) typically abandon their young about 15 days postpartum and produce another litter every 21 days. At high densities, females do not abandon their brood nest, and they allow the young to remain without dispersing. Extended families therefore develop, including the young of several previous litters, the members of which mature more slowly and breed irregularly (Jannett, 1978). In a sense, montane marmots exemplify a similar situation, not only in their reproductive and social patterns but also in the local population density they experience. This density, in turn, occurs because of their delayed dispersal and increased social tolerance, which are associated with coloniality in the first place.

Among free-living mountain goats (*Oreamnos americanus*), yearlings typically leave their mothers and aggregate into peer groups when their mother reproduces. But when nannies experience a barren year, either because they fail to breed or because their kid dies early postpartum, offspring of the previous year are retained by their mothers, establishing a prolonged social relationship that apparently benefits both mother and offspring (Hutchins, 1984).

Correlations between body size and sociality have also been recognized for other taxonomic groups, including the tree shrews (Sorenson, 1970) and the African ungulates (Jarman, 1974). Quality of forage, forage competition, maturation rates, and defense against predators may influence the evolution of sociality in all these groups no less than among the ground squirrels. But as the units of study become increasingly disparate, it seems increasingly important to examine each taxon separately, because the evolutionary factors operating in one may not be readily exportable to another.

A useful cross-taxon parallel may be found, however, in the suggestion of "phenotypic syndromes" (Geist, 1978), in which "dispersal phenotypes," found in situations of high resource abundance, are characterized by high reproductive rates, early maturation, and early dispersal, whereas "maintenance phenotypes," found in situations of resource scarcity, tend to mature and disperse late and to reproduce at a greater age. Woodchucks therefore exemplify dispersal phenotypes, hoary and Olympic marmots exemplify maintenance phenotypes, and yellow-bellied marmots are intermediate, with a tendency to favor one extreme or the other, depending on the habitat. The phenotypic syndrome schema, however, was originally developed for ungulates, and it varies in at least one important respect from the marmot pattern: dispersal-phase ungulates tend to be large and maintenance-phase ungulates tend to be small, whereas among marmots the correlation is precisely the opposite.

There is much more work to be done: nothing is known, for example, about the phenotypic flexibility of marmot social behavior and reproduction. The likely response of individuals presumably adapted to one environmental regime, transplanted into another, remains to be determined. What would become, for example, of woodchucks released at high elevations, or of hoary marmots moved to old fields in Pennsylvania? We are still lacking detailed sociobiologic studies of low-elevation yellow-bellied marmots and of high-latitude woodchucks in Alaska. We might predict adaptations of the woodchuck type among the former and of the hoary/Olympic marmot type among the latter, although with caveats such as the possibility that low-elevation yellow-bellied marmots may experience restricted nutrient availability because of drought and high-latitude woodchucks may inhabit meadows that make up in lushness what they lack in seasonal availability. The wide latitudinal range inhabited by woodchucks also provides the opportunity for examining predicted clines in basic life-history traits; *M. monax* is virtually unstudied, for example, in the extreme southern part of its range as well as in the extreme north.

All things being equal, we can predict adaptations of the *caligata* type among the little-known high-altitude Asian species such as *M. sibirica*, *caudata*, and *himalayana* and for the high-latitude *M. broweri* of Alaska and its Asiatic counterpart, *M. camtschatica*. Adaptations of the *flaviventris* type can be predicted among the steppe species such as *M. bobak* and *baibacina*. And finally, the diminutive, relict *M. menzbieri* would especially repay study, since despite its small size it occupies mountain meadows at high elevations in the western ranges of the USSR's Tien Shan Mountains. Its life-history traits might resemble those of the woodchuck, which it resembles in size, or of the hoary marmot, which it resembles in habitat. And why does a small marmot live in montane situations usually occupied by large-bodied species?

Regardless of the general patterns that characterize marmot sociobiology, and despite the seductiveness of working them out, it is also important to recognize that the pattern of social behavior within any colony or local population of marmots is doubtless modulated by such proximate factors as immediate habitat physiognomy, local kinship situations, age and sex composition at any given time, the day-to-day experiences of colony residents with predators, climate and other varying factors, the individual behavioral profiles of residents during any given year, the number of such individuals, and so forth. On the other hand, regardless of such proximate factors and their undoubted significance, we also should not miss the forest for the trees: there are patterns, both within and between species, and these patterns tell us something about predictability, order, and meaning in the life of marmots—and in life itself.

Summary and Conclusions

1. The marmots reveal numerous correlations among body size, environmental regimes, age at dispersal and sexual maturation, patterns of reproductive effort, and sociality.

2. These correlations are perhaps tied together by the necessity of

large-bodied animals inhabiting severe environments to delay sexual maturation and hence dispersal, as well as by constraints resulting in a cline of reproductive effort from smaller-bodied to larger-bodied marmots.

3. To some extent, these interspecific trends hold intraspecifically as well, especially with regard to life-history characteristics. Among the ground squirrels generally, sociality tends to correlate with large body size, late maturation, late dispersal, and slow reproductive rates.

4. The evolution of sociality among the ground squirrels seems to involve delayed maturation and dispersal most prominently, but also includes the evolution of female/female affiliation and male defense of harems. Body-size energetics and nutrient regimes, kin selection, and sexual selection seem especially important, although other factors such as protection against infanticide may also be relevant.

5. Numerous predictions can be made regarding comparable correlations among other marmot species, especially the Palearctic forms, as well as among Nearctic populations that have not yet been studied. In such cases, however, definitions of "environmental severity" should be treated with caution, since it seems likely to have several potentially significant dimensions.

6. The marmots provide an excellent example of the power of sociobiology to provide correlational insights; there is, however, a lot more to do.

Reference Matter

References Cited

Alexander, R. D., J. Hoogland, R. Howard, K. Noonan, and P. W. Sherman. 1979. Sexual dimorphism and breeding systems in pinnipeds, ungulates, primates and humans. In N. Chagnon and W. Irons (eds.), *Evolutionary Biology and Human Social Behavior: An Anthropological Perspective*. Duxbury Press: North Scituate, Mass.

Altmann, S. A., and J. Altmann. 1977. On the analysis of the rates of behavior. *Anim. Behav.* 25: 364–372.

Andersen, D. C., K. B. Armitage, and R. S. Hoffmann. 1976. Socioecology of marmots: Female reproductive strategies. *Ecology* 57: 552–560.

Anthony, M. 1962. Activity and behavior of the woodchuck in southern Illinois. *Occas. Pap. Adams Ctr. Ecol. Stud.* 6: 1–25.

Armitage, K. B. 1962. Social behavior of a colony of the yellow-bellied marmot (*Marmota flaviventris*). *Anim. Behav.* 10: 319–331.

————. 1965. Vernal behavior of the yellow-bellied marmot. *Anim. Behav.* 13: 59–68.

————. 1973. Population changes and social behavior following colonization by the yellow-bellied marmot. *Jour. Mamm.* 54: 842–858.

————. 1974. Male behavior and territoriality in the yellow-bellied marmot. *Jour. Zool.* 192: 233–265.

————. 1975. Social behavior and population dynamics of marmots. *Oikos* 26: 341–354.

————. 1976. Scent marking by yellow-bellied marmots. *Jour. Mamm.* 57: 583–584.

————. 1977. Social variety in the yellow-bellied marmot: A population behavioral system. *Anim. Behav.* 25: 585–593.

————. 1979. Food selectivity by yellow-bellied marmots. *Jour. Mamm.* 660: 628–629.

————. 1981a. Marmots and coyotes: Behavior of prey and predator. *Jour. Mamm.* 63: 503–505.

———. 1981b. Sociality as a life-history tactic of ground squirrels. *Oecologia* 48: 36–49.

———. 1982. Social dynamics of juvenile marmots: Role of kinship and individual variability. *Behav. Ecol. Sociobiol.* 11: 33–36.

———. 1984. Recruitment in yellow-bellied marmot populations: Kinship, philopatry, and individual variability. In J. Murie and G. Michener (eds.), *The Biology of Ground-Dwelling Squirrels*. Univ. Nebraska Press: Lincoln.

———. 1986a. Individuality, social behavior and reproductive success in yellow-bellied marmots (*Marmota flaviventris*). *Ecology* 67: 1186–1193.

———. 1986b. Marmot polygyny revisited: Determinants of male and female reproductive strategies. In D. I. Rubenstein and R. W. Wrangham (eds.), *Ecology of Social Evolution: Field Studies of Birds and Mammals*. Princeton Univ. Press: Princeton, N.J.

———. 1987. Do female yellow-bellied marmots adjust the sex ratio of their offspring? *Amer. Nat.* 129: 501–509.

Armitage, K. B., and J. F. Downhower. 1974. Demography of yellow-bellied marmot populations. *Ecology* 55: 1233–1245.

Armitage, K. B., J. F. Downhower, and G. E. Svendsen. 1976. Seasonal changes in weights of marmots. *Amer. Midl. Nat.* 96: 36–51.

Armitage, K. B., and D. W. Johns. 1982. Kinship, reproductive strategies and social dynamics of yellow-bellied marmots. *Behav. Ecol. Sociobiol.* 11: 55–63.

Armitage, K. B., D. W. Johns, and D. C. Andersen. 1979. Cannibalism among yellow-bellied marmots. *Jour. Mamm.* 60: 205–207.

Axelrod, R. 1984. *The Evolution of Cooperation*. Basic Books: New York.

Bailey, E. D. 1965a. Seasonal changes in metabolic activity of non-hibernating woodchucks. *Can. Jour. Zool.* 43: 905–909.

———. 1965b. The influence of social interaction and season on weight change in woodchucks. *Jour. Mamm.* 46: 438–445.

Bailey, E. D., and D. E. Davis. 1965. The utilization of body fat during hibernation in woodchucks. *Can. Jour. Zool.* 43: 701–707.

Bailey, V. 1936. The mammals and life zones of Oregon. *No. Amer. Fauna* 55: 1–416.

Baldwin, B. H., B. Tennant, T. J. Reimers, R. G. Cowan, and P. W. Concannon. 1985. Circannual changes in serum testosterone concentrations in adult and yearling woodchucks (*Marmota monax*). *Biol. Reprod.* 32: 804–812.

Balph, D. F. 1984. Spatial and social behavior in a population of Uinta ground squirrels: Interrelations with climate and annual cycle. In J. Murie and G. Michener (eds.), *The Biology of Ground-Dwelling Squirrels*. Univ. Nebraska Press: Lincoln.

Barash, D. P. 1973a. Habitat utilization in three species of mountain mammals. *Jour. Mamm.* 54: 247–250.

———. 1973b. Latitudinal replacement in habitat utilization of mountain mammals. *Jour. Mamm.* 54: 535–536.

———. 1973c. The social biology of the Olympic marmot. *Anim. Behav. Monogr.* 6: 171–249.

————. 1973d. Social variety in the yellow-bellied marmot, *Marmota flaviventris*. *Anim. Behav.* 21: 579–584.

————. 1973e. Territorial and foraging behavior of pika (*Ochotona princeps*) in Montana. *Amer. Midl. Nat.* 89: 202–207.

————. 1974a. The evolution of marmot societies: A general theory. *Science* 185: 415–420.

————. 1974b. Mother-infant relations in captive woodchucks, *Marmota monax*. *Anim. Behav.* 22: 446–448.

————. 1974c. The social behavior of the hoary marmot, *Marmota caligata*. *Anim. Behav.* 22: 257–262.

————. 1975a. Ecology of paternal behavior in the hoary marmot *Marmota caligata*: An evolutionary interpretation. *Jour. Mamm.* 56: 613–618.

————. 1975b. Marmot alarm calling and the question of altruistic behavior. *Amer. Midl. Nat.* 94: 468–470.

————. 1975c. Neighbor recognition in two "solitary" carnivores: The raccoon (*Procyon lotor*) and the red fox (*Vulpes fulva*). *Science* 185: 794–796.

————. 1976a. Pre-hibernation behavior of free-living hoary marmots, *Marmota caligata*. *Jour. Mamm.* 57: 182–185.

————. 1976b. Social behavior and individual differences in free-living alpine marmots (*Marmota marmota*). *Anim. Behav.* 24: 27–35.

————. 1980a. The influence of reproductive status on foraging by hoary marmots *Marmota caligata*. *Behav. Ecol. Sociobiol.* 7: 201–205.

————. 1980b. Predictive sociobiology: Mate selection in damselfishes and brood defense in white-crowned sparrows. In G. Barlow and J. Silverberg (eds.), *Sociobiology: Beyond Nature/Nurture?* Westview Press: Boulder, Colo.

————. 1981. Mate guarding and gallivanting by male hoary marmots. *Behav. Ecol. Sociobiol.* 9: 187–193.

————. 1982. *Sociobiology and Behavior*, 2nd edition. Elsevier: New York.

Beecher, M. D., and I. Beecher. 1979. Sociobiology of bank swallows: Reproductive strategy of the male. *Science* 205: 1282–1285.

Bekoff, M. 1977. Mammalian dispersal and the ontogeny of individual abehavioral phenotypes. *Amer. Nat.* 111: 715–732.

————. 1981. Mammalian sibling interactions: Genes, facilitative environments, and the coefficient of familiarity. In D. Gubernick and P. H. Klopfer (eds.), *Parental Behavior in Mammals*. Plenum Press: New York.

Beltz, A., and E. S. Booth. 1952. Notes on the burrowing and food habits of the Olympic marmot. *Jour. Mamm.* 33: 495–496.

Bibikov, D. I. 1967. *The Mountain Marmots of Central Asia and Kazakhstan*. Nauka: Moscow.

Bibikov, D. I., and S. A. Berendaev. 1978. The Altai marmot. In R. P. Zimina (ed.), *Marmots: Their Distribution and Ecology*. Nauka: Moscow.

Bintz, G. L. 1984. Water balance, water stress, and the evolution of seasonal torpor in ground-dwelling sciurids. In J. Murie and G. Michener (eds.), *The Biology of Ground-Dwelling Squirrels*. Univ. Nebraska Press: Lincoln.

Black, C. C. 1972. Holarctic evolution and dispersal of squirrels (Rodentia: Sciuridae). In T. Dobzhansky, M. K. Hecht, and W. C. Steere (eds.), *Evolutionary
Biology*. Appleton-Century-Crofts: New York.

Bliss, L. C., G. M. Courtin, D. L. Pattie, R. R. Riewe, D. W. A. Whitfield, and
P. Widden. 1973. Arctic tundra ecosystems. *Ann. Rev. Ecol. Syst.* 4: 359–399.

Boag, D. A., and J. O. Murie. 1981. Population ecology of Columbian ground
squirrels in southwestern Alberta. *Can. Jour. Zool.* 59: 2230–2240.

Bopp, P. 1955. Kolonialterritorien bei Murmeltieren. *Rev. Suisse Zool.* 62:
255–261.

Boyce, M. S. 1979. Seasonality and patterns of natural selection for life histories.
Amer. Nat. 114: 569–583.

Brody, A. K., and K. B. Armitage. 1985. The effects of adult removal on dispersal of yearling yellow-bellied marmots (*Marmota flaviventris*). *Can. Jour. Zool.*
63: 2560–2564.

Brody, A. K., and J. Melcher. 1985. Infanticide in yellow-bellied marmots (*Marmota flaviventris*). *Anim. Behav.* 33: 673–674.

Bronson, F. H. 1962. Daily and seasonal activity patterns in woodchucks. *Jour.
Mamm.* 43: 425–426.

———. 1963. Some correlates of interaction rate in natural populations of woodchucks. *Ecology* 44: 637–643.

———. 1964. Agonistic behavior in woodchucks. *Anim. Behav.* 12: 470–478.

Bronson, M. T. 1979. Altitudinal variation in the life history of the golden-
mantled ground squirrel (*Spermophilus lateralis*). *Ecology* 60: 272–279.

Carey, H. V. 1983. Foraging and nutritional ecology of yellow-bellied marmots
in the White Mountains of California. Unpubl. Ph.D. dissertation, Univ.
California, Davis.

———. 1985a. Nutritional ecology of yellow-bellied marmots (*Marmota flaviventris*) in the White Mountains of California [USA]. *Holarct. Ecol.* 8: 259–264.

———. 1985b. The use of foraging areas by yellow-bellied marmots (*Marmota
flaviventris*). *Oikos* 44: 273–279.

Carey, H. V., and P. Moore. 1986. Foraging and predation risk in yellow-bellied
marmots (*Marmota flaviventris*). *Amer. Midl. Nat.* 116: 267–275.

Charnov, E. L., and J. P. Finerty. 1980. Vole population cycles: A case for kin
selection? *Oecologia* 45: 1–2.

Chekalin, V. B. 1965. Ecology of the gray marmot and sanitation of the natural
centre of plague in the north-western part of the central Tien Shan. English
abstract of Ph.D. dissertation, Alma-Ata, USSR.

Christian, J. J. 1970. Social subordination, population density, and mammalian
evolution. *Science* 146: 1550–1560.

Clark, A. B. 1978. Sex ratio and local resource competition in a prosimian primate. *Science* 201: 163–165.

Clutton-Brock, T. H., and P. Harvey. 1983. The functional significance of variation in body size among mammals. In J. F. Eisenberg and D. Kleiman
(eds.), *Advances in the Study of Mammalian Behavior*. Spec. Publ. Amer. Soc.
Mamm. 7.

Cole, L. C. 1954. The population consequences of life history phenomena. *Quart. Rev. Biol.* 29: 103–137.

Couch, L. K. 1930. Notes on the pallid yellow-bellied marmot. *Murrelet* 11: 3–7.

Couturier, M. 1964. *Le Gibier des Montagnes Françaises.* Arthaud: Grenoble.

Davidov, G. S., I. M. Neranov, G. P. Usachev, and E. P. Yakovlev. 1978. The long-tailed marmot. In R. P. Zimina (ed.), *Marmots: Their Distribution and Ecology.* Nauka: Moscow.

Davis, D. E. 1958. The role of density in aggressive behavior of house mice. *Anim. Behav.* 6: 207–210.

———. 1967a. The role of environmental factors in hibernation of woodchucks (*Marmota monax*). *Ecology* 48: 683–689.

———. 1967b. The annual rhythm of fat deposition in woodchucks (*Marmota monax*). *Physiol. Zool.* 40: 391–402.

———. 1971. Annual rhythms in woodchucks (*Marmota monax*). *Ekologiya* 3: 82–87.

———. 1976. Hibernation and circannual rhythms of food consumption in marmots and ground squirrels. *Quart. Rev. Biol.* 51: 477–514.

Davis, L. S. 1984. Behavioral interactions of Richardson's ground squirrels: Asymmetries based on kinship. In J. Murie and G. Michener (eds.), *The Biology of Ground-Dwelling Squirrels.* Univ. Nebraska Press: Lincoln.

Dawkins, R. 1980. Good strategy or evolutionarily stable strategy? In G. Barlow and J. Silverberg (eds.), *Sociobiology: Beyond Nature/Nurture?* Westview Press: Boulder, Colo.

Dawkins, R., and J. Krebs. 1978. Animal signals: Information of manipulation? In J. Krebs and N. Davies (eds.), *Behavioral Ecology.* Sinauer Associates: Sunderland, Mass.

Del Moral, R. 1984. The impact of the Olympic marmot (*Marmota olympus*) on subalpine vegetation structure. *Amer. Jour. Bot.* 71: 1228–1236.

de Vos, A., and D. I. Gillespie. 1960. A study of woodchucks on an Ontario farm. *Can. Field-Nat.* 74: 140–145.

Dobson, F. S. 1979. An experimental study of dispersal in the California ground squirrel. *Ecology* 60: 1103–1109.

———. 1983. Agonism and territoriality in the California ground squirrel. *Jour. Mamm.* 64: 218–225.

———. 1984. Environmental influences on sciurid mating systems. In J. Murie and G. Michener (eds.), *The Biology of Ground-Dwelling Squirrels.* Univ. Nebraska Press: Lincoln.

Dobson, F. S., and J. D. Kjelgaard. 1985a. The influence of food resources on life history in Columbian ground squirrels (*Spermophilus columbianus*). *Can. Jour. Zool.* 63: 2105–2109.

———. 1985b. The influences of food resources on population dynamics in Columbian ground squirrels (*Spermophilus columbianus*). *Can. Jour. Zool.* 63: 2095–2104.

Douglas, G. W., and L. C. Bliss. 1977. Alpine and high subalpine plant communities of the North Cascade range, Washington and British Columbia. *Ecol. Monogr.* 113–150.

Downhower, J. F., and K. B. Armitage. 1971. The yellow-bellied marmot and the evolution of polygamy. *Amer. Nat.* 105: 355–370.

———. 1981. Dispersal of yellow-bellied marmots (*Marmota flaviventris*). *Anim. Behav.* 29: 1064–1069.

Downhower, J. F., and J. D. Pauley. 1970. Automated recordings of body temperature from free-ranging yellow-bellied marmots. *Jour. Wildl. Mgmt.* 34: 639–641.

Dunford, C. 1977a. Kin selection of ground squirrel alarm calls. *Amer. Nat.* 111: 782–785.

———. 1977b. Social system of round-tailed ground squirrels. *Anim. Behav.* 25: 885–906.

Eisenberg, J. F. 1981. *The Mammalian Radiations.* Univ. Chicago Press: Chicago.

Ellerman, J. R., and T. C. S. Morrison-Scott. 1951. *Checklist of Palearctic and Indian Mammals, 1759 to 1946.* British Museum of Natural History: London.

Elliott, P. F. 1975. Longevity and the evolution of polygamy. *Amer. Nat.* 109: 281–287.

Festa-Bianchet, M. 1981. Reproduction in yearling female Columbian ground squirrels (*Spermophilus columbianus*). *Can. Jour. Zool.* 59: 1032–1035.

Festa-Bianchet, M., and D. A. Boag. 1982. Territoriality in adult female Columbian ground squirrels (*Spermophilus columbianus*). *Can. Jour. Zool.* 60: 1060–1066.

Filonov, K. P. 1961. Data on the ecology of *Marmota camtschatica* in the Barguzin state forest. *Trudy Barguzinsk. Gosud. Zapov.* 3: 169–180.

Fisher, R. A. 1958. *The Genetical Theory of Natural Selection.* Dover: New York.

Foltz, D. W., and J. L. Hoogland. 1981. Analysis of the mating system in the black-tailed prairie dog (*Cynomys ludovicianus*) by likelihood of paternity. *Jour. Mamm.* 62: 706–712.

Formozov, A. N. 1966. Adaptive modifications of behavior in mammals of the Eurasian steppes. *Jour. Mamm.* 47: 208–233.

Frase, B. A., and K. B. Armitage. 1984. Foraging patterns of yellow-bellied marmots (*Marmota flaviventris*): Role of kinship and individual variability. *Behav. Ecol. Sociobiol.* 16: 1–10.

Freeland, W. J., and D. H. Janzen. 1974. Strategies in herbivory by mammals: The role of plant secondary compounds. *Amer. Nat.* 108: 269–289.

French, A. R. 1982. Intraspecific differences in the pattern of hibernation in the ground squirrel, *Spermophilus beldingi. Jour. Comp. Phys.* 148: 83–91.

Fretwell, S. D., and H. L. Lucas. 1969. On territorial behavior and other factors influencing habitat distribution in birds. *Acta Biotheor.* 19: 16–36.

Garrott, R. A., and D. A. Jenni. 1978. Arboreal behavior of yellow-bellied marmots. *Jour. Mamm.* 59: 433–434.

Geist, V. 1978. *Life Strategies, Human Evolution, Environmental Design.* Springer-Verlag: New York.

Green, J. E. 1977. Population regulation and annual cycles of activity and dispersal in the Arctic ground squirrel. Unpubl. M.S. thesis, Univ. British Columbia, Vancouver.

Grizzell, R. A. 1955. A study of the southern woodchuck, *Marmota monax monax*. *Amer. Midl. Nat.* 53: 257–293.

Gross, M., and E. L. Charnov. 1980. Alternative male life histories in bluegill sunfish. *Proc. Nat. Acad. Sci.* 77: 6937–6940.

Hafner, D. J. 1984. Evolutionary relationships of the nearctic Sciuridae. In J. Murie and G. Michener (eds.), *The Biology of Ground-Dwelling Squirrels*. Univ. Nebraska Press: Lincoln.

Haggerty, S. M. 1968. The ecology of the Franklin's ground squirrel (*Citellus franklinii*) at Itasca State Park, Minn. Unpubl. M.S. thesis, Illinois State Univ., Normal.

Halpin, Z. T. 1984. The role of olfactory communication in the social systems of ground-dwelling sciurids. In J. Murie and G. Michener (eds.), *The Biology of Ground-Dwelling Squirrels*. Univ. Nebraska Press: Lincoln.

Hamilton, W. D. 1964. The genetical theory of social behavior. *Jour. Theor. Biol.* 7: 1–52.

———. 1971. Geometry for the selfish herd. *Jour. Theor. Biol.* 31: 295–311.

Hamilton, W. J. 1934. The life history of the rufescent woodchuck. *Ann. Carnegie Mus.* 23: 87–118.

Hanken, J., and P. W. Sherman. 1981. Multiple paternity in Belding's ground squirrel litters. *Science* 212: 351–353.

Hansen, R. M. 1975. Foods of the hoary marmot on Kenai Peninsula, Alaska. *Jour. Mamm.* 94: 348–353.

Haslett, G. W. 1973. The significance of anal scent marking in the eastern woodchuck. *Bull. Ecol. Soc. Amer.* 54: 43–44.

Heard, D. C. 1977. The behaviour of Vancouver Island marmots, *Marmota vancouverensis*. Unpubl. M.S. thesis, Univ. British Columbia, Vancouver.

Hirsch, J. 1963. Behavior genetics and individuality understood. *Science* 142: 1436–1442.

———. 1967. Behavior-genetic analysis. In J. Hirsch (ed.), *Behavior-Genetic Analysis*. McGraw-Hill: New York.

Hock, R. J. 1969. Thermoregulatory variations of high-altitude hibernators in relation to ambient temperature, season, and hibernation. *Fed. Proc.* 28: 1047–1052.

Hofer, S., and P. Ingold. 1984. The whistles of the alpine marmot (*Marmota marmota marmota*): Their structure and occurrence in the antipredator context. *Rev. Suisse Zool.* 91: 861–866.

Hoffmann, R. S. 1974. Terrestrial vertebrates. In J. Ives and R. Barry (eds.), *Arctic and Alpine Environments*. Methuen: London.

Hoffmann, R. S., J. W. Koeppl, and C. F. Nadler. 1979. The relationships of the amphiberingian marmots (Mammalia: Sciuridae). *Occas. Pap. Mus. Nat. Hist. Univ. Kan.* 83: 1–56.

Hoffmann, R. S., and C. F. Nadler. 1968. Chromosomes and systematics of some North American species of the genus *Marmota* (Rodentia: Sciuridae). *Experientia* 24: 740–742.

Holecamp, K. E. 1983. Proximal mechanisms of natal dispersal in Belding's ground squirrels (*Spermophilus beldingi*). Unpubl. Ph.D. dissertation, Univ. California, Berkeley.

———. 1984. Dispersal in ground-dwelling sciurids. In J. Murie and G. Michener (eds.), *The Biology of Ground-Dwelling Squirrels*. Univ. Nebraska Press: Lincoln.

———. 1985. Natal dispersal in Belding's ground squirrels (*Spermophilus beldingi*). *Behav. Ecol. Sociobiol.* 16: 21–30.

———. 1986. Proximal causes of natal dispersal in Belding's ground squirrels (*Spermophilus beldingi*). *Ecol. Monogr.* 56: 365–392.

Holmes, W. G. 1979. Social behavior and foraging strategies of hoary marmots (*Marmota caligata*) in Alaska. Unpubl. Ph.D. dissertation, Univ. Washington, Seattle.

———. 1984a. Predation risk and foraging behavior of the hoary marmot in Alaska. *Behav. Ecol. Sociobiol.* 15: 293–301.

———. 1984b. The ecological basis of monogamy in Alaskan hoary marmots. In J. Murie and G. Michener (eds.), *The Biology of Ground-Dwelling Squirrels*. Univ. Nebraska Press: Lincoln.

———. 1986. Identification of paternal half-siblings by captive Belding's ground squirrels (*Spermophilus beldingi*). *Anim. Behav.* 34: 321–327.

Holmes, W. G., and P. W. Sherman. 1982. The ontogeny of kin recognition in two species of ground squirrels. *Amer. Zool.* 22: 491–517.

Hoogland, J. L. 1979. Aggression, ectoparasitism, and other possible costs of prairie dog (Sciuridae, *Cynomys* spp.) coloniality. *Behaviour* 69: 1–35.

———. 1981a. The evolution of coloniality in white-tailed and black-tailed prairie dogs. *Ecology* 62: 252–272.

———. 1981b. Nepotism and cooperative breeding in the black-tailed prairie dogs (Sciuridae: *Cynomys leucurus* and *C. ludovicianus*). In R. D. Alexander and D. W. Tinkle (eds.), *Natural Selection and Social Behavior*. Chiron Press: New York.

———. 1982. Prairie dogs avoid extreme inbreeding. *Science* 215: 1639–1641.

———. 1983. Black-tailed prairie dog coteries are cooperatively breeding units. *Amer. Nat.* 121: 275–280.

———. 1985. Infanticide in prairie dogs (*Cynomys ludovicianus*): Lactating females kill offspring of close kin. *Science* 230: 1037–1040.

———. 1986. Infanticide in prairie dogs (*Cynomys ludovicianus*) varies with competition but not with kinship. *Anim. Behav.* 34: 263–270.

Howell, A. H. 1915. Revision of the American marmots. U.S. Government Printing Office: Washington, D.C.

Hutchins, M. D. 1984. Mother-offspring relationships in mountain goats (*Oreamnos americanus*). Unpubl. Ph.D. dissertation, Univ. Washington, Seattle.

Ismaghilov, M. I. 1956. Contribution to the ecology of *Marmota bobak centralis* Thomas. *Zool. Zhur.* 35: 11.

Ives, J. D. 1974. Biological refugia and the nunatak hypothesis. In J. D. Ives and R. G. Barry (eds.), *Arctic and Alpine Environments*. William Clowes and Sons: London.

Jamieson, S. H., and K. B. Armitage. 1987. Sex differences in the play behavior of yearling yellow-bellied marmots. *Ethology* 74: 237–253.

Janis, C. 1976. The evolutionary strategy of the Equidae and the origins of rumen and cecal digestion. *Evolution* 30: 757–774.

Jannett, F. J. 1978. The density-dependent formation of extended maternal families of the montane vole, *Microtus montanus manus*. *Behav. Ecol. Sociobiol.* 3: 245–263.

Jarman, P. J. 1974. The social organization of antelope in relation to their ecology. *Behaviour* 48: 215–266.

Johns, D. W., and K. B. Armitage. 1979. Behavioral ecology of alpine yellow-bellied marmots. *Behav. Ecol. Sociobiol.* 5: 133–157.

Kalabukhov, N. I. 1960. Comparative ecology of hibernating mammals. *Bull. Mus. Comp. Zool.* 124: 45–74.

Kapitonov, V. I. 1960. A sketch on biology of the black-capped marmot. *Zool. Zhur.* 39: 448–457.

———. 1978. The Kamchatka marmot. In R. P. Zimina (ed.), *Marmots: Their Distribution and Ecology*. Nauka: Moscow.

Kiell, D. F., and J. S. Millar. 1978. Growth of juvenile Arctic ground squirrels (*Spermophilus parryi*) at McConnell River, NWT. *Can. Jour. Zool.* 56: 1475–1478.

Kilgore, D. L. 1972. Energy dynamics of the yellow-bellied marmot (*Marmota flaviventris*): A hibernator. Unpubl. Ph.D. dissertation, Univ. Kansas, Lawrence.

Kilgore, D. L., and K. B. Armitage. 1978. Energetics of yellow-bellied marmot populations. *Ecology* 59: 78–88.

King, W. J., and J. O. Murie. 1985. Temporal overlap of female kin in Columbian ground squirrels (*Spermophilus columbianus*). *Behav. Ecol. Sociobiol.* 16: 337–342.

King's College Sociobiology Group. 1982. *Current Problems in Sociobiology*. Cambridge Univ. Press: Cambridge, England.

Kivett, V. K., J. O. Murie, and A. L Steiner. 1976. A comparative study of scent-gland location and related behavior in some northwestern Nearctic ground squirrel species (Sciuridae): An evolutionary approach. *Can. Jour. Zool.* 54: 1294–1306.

Kizilov, V. A., and S. A. Berendaev. 1978. The long-tailed marmot. In R. P. Zimina (ed.), *Marmots: Their Distribution and Ecology*. Nauka: Moscow.

Kizilov, V. A., and N. I. Semenova. 1967. On winter hibernation of the red marmot. In *Research on the Mammal Fauna of the USSR*. Nauka: Moscow.

Kleiman, D. G., and J. J. Eisenberg. 1973. Comparisons of canid and felid social systems from an evolutionary perspective. *Anim. Behav.* 21: 637–659.

Koenig, L. 1957. Beobachtungen über Reviermarkierung sowie Droh- Kampf- und Abwehrverhalten des Murmeltieres (*Marmota marmota* L.). *Zeit. Tierpsychol.* 14: 510–521.

Koford, C. B. 1958. Prairie dogs, whitefaces, and blue grama. *Wildl. Monogr.* 3: 1–78.

Krebs, J., and N. B. Davies (eds.). 1984. *Behavioral Ecology.* Sinauer Associates: Sunderland, Mass.

Kuramoto, R. T., and L. C. Bliss. 1970. Ecology of subalpine meadows in the Olympic Mountains, Washington. *Ecol. Monogr.* 40: 317–347.

Kurland, J. 1977. Kin selection in the Japanese monkey. *Contributions to Primatology*, Vol. 35. Karger: Basel.

Langenheim, J. H. 1956. Plant succession on a subalpine earth-flow in Colorado. *Ecology* 37: 301–317.

Lloyd, J. E. 1972. Vocalizations in *Marmota monax*. *Jour. Mamm.* 53: 214–216.

Loibl, M. F. 1983. Social and non-social behaviors of arctic marmots, *Marmota broweri*, housed outdoors. Unpubl. M.S. thesis, Univ. Maryland, College Park.

Lyman, C. P., and A. R. Dawe (eds.). 1960. Mammalian hibernation. *Bull. Mus. Comp. Zool.* 124: 1–549.

Lyman, C. P., J. S. Willis, A. Malan, and L. C. H. Wang. 1982. *Hibernation and Torpor in Mammals and Birds.* Academic Press: New York.

Marler, P. 1957. Specific distinctiveness in the communication signals of birds. *Behaviour* 11: 13–39.

Marr, N. V., and R. L. Knight. 1983. Food habits of golden eagles in eastern Washington. *Murrelet* 64: 73–77.

Martell, A. M., and R. J. Milko. 1986. Seasonal diets of Vancouver Island (Canada) marmots, *Marmota vancouverensis*. *Can. Field-Nat.* 100: 241–245.

Maschkin, V. I. 1982. New materials on ecology of *Marmota menzbieri*. *Zool. Zhur.* 61: 278–289.

Maxwell, C. S., and M. L. Morton. 1975. Comparative thermoregulatory capabilities of neonatal ground squirrels. *Jour. Mamm.* 56: 821–828.

Mayer, W. V., and E. T. Roche. 1954. Developmental patterns in the Barrow ground squirrel (*Citellus parryi barrowensis*). *Growth* 18: 53–69.

Maynard Smith, J. 1964. Group selection and kin selection. *Nature* 201: 1145–1147.

———. 1976. Evolution and the theory of games. *Amer. Scientist* 64: 41–45.

———. 1982. *Evolution and the Theory of Games.* Cambridge Univ. Press: New York.

Mayr, E. 1963. *Animal Species and Evolution.* Harvard Univ. Press: Cambridge, Mass.

McLean, I. G. 1982. The association of female kin in the arctic ground squirrel. *Behav. Ecol. Sociobiol.* 10: 91–99.

———. 1983. Parental behavior and killing of young in arctic ground squirrels. *Anim. Behav.* 31: 32–44.

———. 1984. Spacing behavior and aggression in female ground squirrels. In J. Murie and G. Michener (eds.), *The Biology of Ground-Dwelling Squirrels.* Univ. Nebraska Press: Lincoln.

McNab, B. N. 1971. On the ecological significance of Bergmann's Rule. *Ecology* 52: 845–854.

Meier, P. T. 1985. Behavioral ecology, social organization and mating system of woodchucks (*Marmota monax*) in southeast Ohio. Unpubl. Ph.D. dissertation, Ohio University, Athens.

Merriam, H. G. 1971. Woodchuck burrow distribution and related movement patterns. *Jour. Mamm.* 52: 732–746.

Michener, G. R. 1973. Intraspecific aggression and social organization in ground squirrels. *Jour. Mamm.* 54: 1001–1003.

———. 1979. Spatial relationships and social organization of adult Richardson's ground squirrels. *Can. Jour. Zool.* 57: 125–139.

———. 1983. Kin identification, matriarchies, and the evolution of sociality in ground-dwelling sciurids. In J. F. Eisenberg and D. G. Kleiman (eds.), *Recent Advances in the Study of Mammalian Behavior.* Spec. Publ. Amer. Soc. Mamm. 7.

———. 1984. Age, sex and species differences in the annual cycles of ground-dwelling sciurids: Implications for sociality. In J. Murie and G. Michener (eds.), *The Biology of Ground-Dwelling Squirrels.* Univ. Nebraska Press: Lincoln.

Michener, G. R., and D. R. Michener. 1973. Spatial distribution of yearlings in a Richardson's ground squirrel population. *Ecology* 54: 1138–1142.

———. 1977. Population structure and dispersal in Richardson's ground squirrels. *Ecology* 58: 359–368.

Michener, G. R., and D. H. Sheppard. 1972. Social behavior between adult female Richardson's ground squirrels (*Spermophilus richardsonii*) and their own and alien young. *Can. Jour. Zool.* 50: 1343–1349.

Milko, R. J. 1984. Vegetation and foraging ecology of the Vancouver Island marmot (*Marmota vancouverensis*). Unpubl. M.S. thesis, Univ. Victoria, Victoria, B.C.

Morrison, P. 1960. Some interrelationships between weight and hibernation function. *Bull. Mus. Comp. Zool.* 124: 75–90.

Muller-Using, D. 1954. Beitrage zur Oekologie der *Marmota marmota marmota. Zeit. Saugetier.* 19: 166–177.

———. 1955. Vom "Pfeifen" des Murmeltieres. *Zeit. Jagdwissen.* 1: 32–33.

———. 1956. Zum Verhalten des Murmeltieres (*Marmota marmota*). *Zeit. Tierpsychol.* 13: 135–142.

———. 1957. Die Paarungsbiologie des Murmeltieres. *Zeit. Jagdwissen.* 3: 24–28.

Munch, H. 1958. Zur Oekologie und Psychologie von *Marmota m. marmota. Zeit. Saugetier.* 23: 129–138.

Murie, J. O. 1973. Population characteristics and phenology of a Franklin ground squirrel (*Spermophilus franklinii*) colony in Alberta, Canada. *Amer. Midl. Nat.* 90: 334–340.

Murie, J. O., and M. A. Harris. 1978. Territoriality and dominance in male Columbian ground squirrels (*Spermophilus columbianus*). *Can. Jour. Zool.* 56: 2404–2412.

———. 1982. Annual variation of spring emergence and breeding in Columbian ground squirrels (*Spermophilus columbianus*). *Jour. Mamm.* 63: 431–439.

Nee, J. A. 1969. Reproduction in a population of yellow-bellied marmots (*Marmota flaviventris*). *Jour. Mamm.* 50: 7556–7565.

Nekipelov, N. V. 1978. The tarbagan in southeast Zabaikal. In R. P. Zimina (ed.), *Marmots: Their Distribution and Ecology*. Nauka: Moscow.

Nikolskii, A. A. 1974. Geographical variability of sound calls and rhythmic organization in marmots of the group bobak. *Zool. Zhur.* 53: 436–444.

Novikov, G. A. 1956. *Carnivorous Mammals of the Fauna of the USSR*. Nauka: Moscow. [English translation, 1962, Israel Program of Scientific Translations: Jerusalem.]

Nowicki, S., and K. B. Armitage. 1979. Behavior of juvenile yellow-bellied marmots: Play and social integration. *Zeit. Tierpsychol.* 51: 85–105.

Noyes, D. H., and W. G. Holmes. 1979. Behavioral responses of free-living hoary marmots to a model golden eagle. *Jour. Mamm.* 60: 408–411.

Orians, G. H. 1969. On the evolution of mating systems in birds and mammals. *Amer. Nat.* 103: 589–603.

Owings, D. H., and D. F. Hennessey. 1984. The importance of variation in sciurid visual and vocal communication. In J. Murie and G. Michener (eds.), *The Biology of Ground-Dwelling Squirrels*. Univ. Nebraska Press: Lincoln.

Pattie, D. L. 1967. Observations on an alpine population of yellow-bellied marmots (*Marmota flaviventris*). *Northwest Sci.* 41: 96–102.

Pfeifer, S. R. 1980. Demographic and behavioral influences on juvenile Wyoming ground squirrel dispersal. Unpubl. Ph.D. dissertation, Univ. Colorado, Boulder.

———. 1982. Disappearance and dispersal of *Spermophilus elegans* juveniles in relation to behavior. *Behav. Ecol. Sociobiol.* 10: 237–243.

Phillips, J. A. 1984. Environmental influences on reproduction in the golden-mantled ground squirrel. In J. Murie and G. Michener (eds.), *The Biology of Ground-Dwelling Squirrels*. Univ. Nebraska Press: Lincoln.

Ralls, K. 1976. Mammals in which females are larger than males. *Quart. Rev. Biol.* 51: 245–276.

Rausch, R. L. 1953. On the status of some arctic mammals. *Arctic* 6: 91–148.

Rausch, R. L., and V. R. Rausch. 1971. The somatic chromosomes of some North American marmots (Sciuridae), with remarks on the relationships of *Marmota broweri* Hall and Gilmore. *Mammalia* 35: 85–101.

Raynor, L. S. 1985. Effects of habitat quality on growth, age of first reproduction, and dispersal in Gunnison's prairie dogs (*Cynomys gunnisoni*). *Can. Jour. Zool.* 63: 2835–2840.

Reichman, F. 1942. Groundhog day. *Amer. Germ. Rev.* Feb. 8(3): 11–13.

Reynolds, H. G., and F. Turkowski. 1972. Reproductive variations in the round-tailed ground squirrel as related to winter rainfall. *Jour. Mamm.* 53: 893–898.

Rohwer, S., S. Fretwell, and D. Niles. 1980. Delayed maturation in passerine plumages and the deceptive acquisition of resources. *Amer. Nat.* 115: 400–437.

Rubenstein, D. I. 1980. On the evolution of alternative mating strategies. In J. Staddon (ed.), *Limits to Action*. Academic Press: New York.

Rubenstein, D. I., and R. W. Wrangham. 1980. Why is altruism towards kin so rare? *Zeit. Tierpsychol.* 54: 381–387.

Schaller, G. B. 1977. *Mountain Monarchs*. Univ. Chicago Press: Chicago.

Schmutz, S. M., D. A. Boag, and J. K. Schmutz. 1979. Causes of the unequal sex ratio in populations of adult Richardson's ground squirrels. *Can. Jour. Zool.* 57: 1849–1855.

Scholander, P. F. 1955. Evolution of climatic adaptation in homeotherms. *Evolution* 9: 15–26.

Schwagmeyer, P. L. 1980. Alarm calling behavior of thirteen-lined ground squirrels, *Spermophilus tridecemlineatus. Behav. Ecol. Sociobiol.* 7: 195–200.

———. 1984. Multiple mating and intersexual selection in thirteen-lined ground squirrels. In J. Murie and G. Michener (eds.), *The Biology of Ground-Dwelling Squirrels*. Univ. Nebraska Press: Lincoln.

Schwagmeyer, P. L., and C. H. Brown. 1983. Factors affecting male-male competition in thirteen-lined ground squirrels. *Behav. Ecol. Sociobiol.* 13: 1–6.

Schwagmeyer, P. L., and S. J. Woontner. 1985. Mating competition in an asocial ground squirrel, *Spermophilus tridecemlineatus. Behav. Ecol. Sociobiol.* 17: 291–296.

———. 1986. Scramble competition polygyny in thirteen-lined ground squirrels (*Spermophilus tridecemlineatus*): The relative contributions of overt conflict and competitive mate searching. *Behav. Ecol. Sociobiol.* 19: 359–364.

Schwartz, O. A., and K. B. Armitage. 1980. Genetic variation in social mammals: The marmot model. *Science* 207: 664–667.

Scott, T. C., and W. D. Klimstra. 1955. Red foxes and a declining prey population. *Monogr. Ser. So. Ill. Univ.* 1.

Searcy, W. A. 1980. Optimum body sizes and different ambient temperatures: An energetics explanation of Bergmann's Rule. *Jour. Theor. Biol.* 83: 579–593.

Sheppard, D. H. 1972. Reproduction of Richardson's ground squirrel (*Spermophilus richardsonii*) in southern Saskatchewan. *Can. Jour. Zool.* 50: 1577–1581.

Sherman, P. W. 1976. Natural selection among some group-living organisms. Unpubl. Ph.D. dissertation, Univ. Michigan, Ann Arbor.

———. 1977. Nepotism and the evolution of alarm calls. *Science* 197: 1246–1253.

———. 1980. The limits of ground squirrel nepotism. In G. Barlow and J. Silverberg (eds.), *Sociobiology: Beyond Nature/Nurture?* Westview Press: Boulder, Colo.

———. 1981. Reproductive competition and infanticide in Belding's ground squirrels and other organisms. In R. D. Alexander and D. W. Tinkle (eds.), *Natural Selection and Social Behavior*. Chiron Press: New York.

———. 1985. Alarm calls of Belding's ground squirrel (*Spermophilus beldingi*) to aerial predators: Nepotism or self-preservation? *Behav. Ecol. Sociobiol.* 17: 313–324.

Sherman, P. W., and M. L. Morton. 1984. Demography of Belding's ground squirrels. *Ecology* 65: 1617–1628.

Shirer, H. W., and J. F. Downhower. 1968. Radio-tracking of dispersing yellow-bellied marmots. *Trans. Kan. Acad. Sci.* 71: 463–479.

Shubin, I. G. 1962. The reproduction times of *Marmota bobak*. *Zool. Zhur.* 42: 274–281.

———. 1963. Hibernation times of steppe marmot and *Citellus pygmaeus* in Central Kazakhstan. *Zool. Zhur.* 42: 274–281.

Shubin, I. G., V. I. Abelentsev, and S. N. Semikhatova. 1978. The steppe marmot. In R. P. Zimina (ed.), *Marmots: Their Distribution and Ecology*. Nauka: Moscow.

Slade, N. A., and D. F. Balph. 1974. Population ecology of Uinta ground squirrels. *Ecology* 55: 989–1003.

Smirin, Y. M., N. A. Formozov, D. I. Bibikov, and D. Myagmarzhav. 1985. Characteristics of two marmot species (Mammalia, Rodentia, Sciuridae) in the zone of their contact in the Mongolian Altai. *Zool. Zhur.* 64: 1873–1885.

Snyder, R. L. 1976. *The Biology of Population Growth*. St. Martin's Press: New York.

Snyder, R. L., and J. J. Christian. 1960. Reproductive cycle and litter size of the woodchuck. *Ecology* 41: 647–656.

Snyder, R. L., D. E. Davis, and J. J. Christian. 1961. Seasonal changes in the weights of woodchucks. *Jour. Mamm.* 42: 297–312.

Sorenson, M. W. 1970. Behavior of tree shrews. In L. A. Rosenblum (ed.), *Primate Behavior*. Academic Press: New York.

Svendsen, G. 1974. Behavioral and environmental factors in the spatial distribution and population dynamics of a yellow-bellied marmot population. *Ecology* 55: 760–771.

———. 1976. Structure and location of burrows of yellow-bellied marmot. *Southwest. Nat.* 20: 487–494.

Svendsen, G., and K. B. Armitage. 1973. Mirror-image stimulation applied to behavioral studies. *Ecology* 54: 623–627.

Taulman, J. F. 1977. Vocalizations of the hoary marmot, *Marmota caligata*. *Jour. Mamm.* 58: 681–683.

Travis, S., and K. B. Armitage. 1972. Some quantitative aspects of the behavior of marmots. *Trans. Kan. Acad. Sci.* 75: 308–321.

Trivers, R. L. 1971. The evolution of reciprocal altruism. *Quart. Rev. Biol.* 46: 35–57.

———. 1972. Parental investment and sexual selection. In B. Campbell (ed.), *Sexual Selection and the Descent of Man*. Aldine: Chicago.

———. 1974. Parent-offspring conflict. *Amer. Zool.* 14: 249–264.

Trivers, R. L., and D. E. Willard. 1973. Natural selection of parental ability to vary the sex ratio of offspring. *Science* 179: 90–92.

Tyser, R. W. 1980. Use of substrate for surveillance behaviors in a community of talus slope mammals. *Amer. Midl. Nat.* 104: 32–38.

Vestal, B. M., and H. McCarley. 1984. Spatial and social relations of kin in thirteen-lined and other ground squirrels. In J. Murie and G. Michener (eds.), *The Biology of Ground-Dwelling Squirrels.* Univ. Nebraska Press: Lincoln.

vom Saal, F. S. 1981. Variation in phenotype due to random intrauterine positioning of male and female fetuses in rodents. *Jour. Reprod. Fert.* 62: 633–650.

Ward, J. M., and K. B. Armitage. 1981a. Circannual rhythms of food consumption, body mass, and metabolism in yellow-bellied marmots. *Comp. Biochem. Phys.* 69: 621–626.

———. 1981b. Water budgets of montane-mesic and lowland-xeric populations of yellow-bellied marmots. *Comp. Biochem. Phys.* 69: 627–630.

Waring, G. H. 1966. Sounds and communications of the yellow-bellied marmot (*Marmota flaviventris*). *Anim. Behav.* 14: 177–183.

Wasser, S. K. (ed.). 1983. *Social Behavior among Female Vertebrates.* Academic Press: New York.

Wasser, S. K., and D. P. Barash. 1983. Reproductive suppression among female mammals: Implications for biomedicine and sexual selection theory. *Quart. Rev. Biol.* 58: 513–538.

Waterman, J. M. 1985. Behavior and use of space by juvenile Columbian ground squirrels (*Spermophilus columbianus*). *Can. Jour. Zool.* 64: 1121–1127.

Webb, D. R. 1980. Environmental harshness, heat stress, and *Marmota flaviventris*. *Oecologia* 44: 390–395.

———. 1981. Macro-habitat patch structure, environmental harshness and *Marmota flaviventris*. *Behav. Ecol. Sociobiol.* 8: 175–182.

Westoby, M. 1974. An analysis of diet selection by large generalist herbivores. *Amer. Nat.* 108: 290–304.

Wiley, R. H. 1974. Evolution of social organization and life history patterns among grouse: Tetraonidae. *Quart. Rev. Biol.* 49: 201–227.

Williams, G. C. 1966. *Adaptation and Natural Selection.* Princeton Univ. Press: Princeton, N.J.

———. 1979. The question of adaptive sex ratio in outcrossed vertebrates. *Proc. Roy. Soc. London* 205: 567–580.

Wilson, E. O. 1975. *Sociobiology: The New Synthesis.* Harvard Univ. Press: Cambridge, Mass.

Wistrand, H. 1974. Individual, social and seasonal behavior of the thirteen-lined ground squirrel (*Spermophilus tridecemlineatus*). *Jour. Mamm.* 55: 329–347.

Wood, W. A. 1973. Habitat selection and energetics of the Olympic marmot. Unpubl. M.S. thesis, Western Washington Univ., Bellingham.

Wright, J., B. C. Tennant, and B. May. 1987. Genetic variation between woodchuck populations with high and low prevalence rates of woodchuck hepatitus virus infection. *Jour. Wildl. Dis.* 23: 186–191.

Wynne-Edwards, V. C. 1962. *Animal Dispersion in Relation to Social Behavior.* Hafner: New York.

Yeaton, R. 1972. Social behavior and social organization in Richardson's ground squirrel (*Spermophilus richardsonii*) in Saskatchewan. *Jour. Mamm.* 55: 139–147.

Zharov, V. R. 1972. Stock and territorial structure of colonies of Camtschatica
    marmots (*Marmota camtschatica*) on the Baruzinsky mountain ridge. *Zool.
    Zhur.* 51: 1387–1394.

Zimina, R. P. (ed.) 1978. *Marmots: Their Distribution and Ecology.* Nauka: Moscow.

Zimina, R. P., and I. P. Gerasimov. 1973. The periglacial expansion of mar-
    mots (*Marmota*) in middle Europe during late Pleistocene. *Jour. Mamm.* 54:
    327–340.

Index

Note: much of this book involves comparisons among the four marmot species *Marmota caligata, flaviventris, monax,* and *olympus.* Because references to these species appear on virtually every page, they are not indexed by name. Other marmot species, however, are included, under the species epithet.

Vancouver marmot bidding the reader
a fond adieu. (Photo by G. W. Smith)

Library of Congress Cataloging-in-Publication Data

Barash, David P.
  Marmots : social behavior and ecology / David P. Barash.
    p.   cm.
  Bibliography: p.
  Includes index.
  ISBN 0-8047-1534-3 (alk. paper)
    1. Marmots—Behavior.   2. Marmots—Ecology.
3. Social behavior in animals.   I. Title.
QL737.R638B37   1989
599.32'32—dc19   89-4284
                                                    CIP